Energy Security

Energy Security

Re-Measuring the World

Sascha Müller-Kraenner

London • Sterling, VA

First published by Earthscan in the UK and USA in 2008
Originally published in 2007 by Verlag Antje Kunstmann GmbH as *Energiesicherheit*

ISBN: 978-1-84407-582-9
Typeset by FiSH Books, Enfield, Middx.
Printed and bound in the UK by TJ International Ltd, Padstow
Cover design by Ruth Bateson
Translated by Intelligent Renewable Energy, Freiburg, Germany
Translator: Gabriele Hase

For a full list of publications please contact:

Earthscan
Dunstan House
14a St Cross St
London, EC1N 8XA, UK
Tel: +44 (0)20 7841 1930
Fax: +44 (0)20 7242 1474
Email: earthinfo@earthscan.co.uk
Web: **www.earthscan.co.uk**

22883 Quicksilver Drive, Sterling, VA 20166-2012, USA

Earthscan publishes in association with the International Institute for Environment and Development

A catalogue record for this book is available from the British Library

Library of Congress Cataloging-in-Publication Data
Müller-Kraenner, Sascha.
Energy security : re-measuring the world / by Sascha Müller-Kraenner.
p. cm.
Includes bibliographical references.
ISBN 978-1-84407-582-9 (hardback)
1. Power Resources. 2. Energy policy. 3. Energy conservation. 4. Energy consumption–Environmental aspects. I. Title.
TJ163.2.M843 2008
333.79–dc22

2008018424

The paper used for this book is FSC-certified and totally chlorine-free. FSC (the Forest Stewardship Council) is an international network to promote responsible management of the world's forests.

Contents

Acknowledgements

I wish to thank many friends and colleagues who helped me in writing this book with their ideas and comments. They are Marc Berthold, Maria Ivanova, Mi-Hyung Kim, R. Andreas Kraemer, Ikuko Matsumoto, Mika Obayashi, Olena Prystayko, Danyel Reiche, Andreas Wagner, Claude Weinber and many more. Seunghee Ham helped me with the background research on Russian and Asian energy policy.

Many of the ideas presented in this book were initially published in the magazine *Kommune*, for which I have written regular columns since the end of 1998. I wish to thank in, particular, the editor in chief of *Kommune*, Michael Ackermann, who helped me establish the first contacts with the publishers. This book would not have been possible without the professional, meticulous and always encouraging assistance of Susanne Eversmann from Antje Kunstmann publishing house. Similar thanks go to Michael Fell from Earthscan for guiding me through the production process of the English language version, as well as to Rian van Staden and his team of translators for a marvellous job.

My special thanks go to the Yale World Fellowship Program that made it possible for me to reflect on the topic of this book for four months. In particular, I wish to thank Daniel Esty and Kel Ginsberg.

And most of all, I am grateful to my wife, Karin Holl, not only for critically reading through the chapters, but for supporting me in every imaginable way, and to Alexander and Nikolas Holl for many helpful tips and encouraging words.

I wish to dedicate this book to Christiane Knospe. Unfortunately, you cannot hold it in your hands any more.

List of Acronyms and Abbreviations

ANWR	Arctic National Wildlife Refuge
ASEAN	Association of South East Asian Nations
ASPO	Association for the Study of Peak Oil and Gas
BfN	Bundesamt für Naturschutz (German Federal Nature Conservation Agency)
BTC	Baku-Tblisi-Ceyhan (pipeline)
CCS	carbon capture and storage
CDU	Christian Democratic Union
CFSP	Common Foreign and Security Policy
CNOOC	China National Offshore Oil Company
CNPC	China National Petroleum Corporation
CPD	Committee on the Present Danger
EBRD	European Bank for Reconstruction and Development
ECSC	European Coal and Steel Community
EEC	European Economic Community
EEZ	Exclusive Economic Zone
EIB	European Investment Bank
EITI	Extractive Industries Transparency Initiative
ESI	Environmental Sustainability Index
ESS	European Security Strategy
EU	European Union
EURATOM	European Atomic Energy Community
GCC	Gulf Cooperation Council
GDR	German Democratic Republic (former East Germany)
HCU	hard coal units
IAEA	International Atomic Energy Agency
IASC	International Arctic Science Committee
IEA	International Energy Agency
IMF	International Monetary Fund
IPCC	Intergovernmental Panel on Climate Change
IRENA	International Renewable Energy Agency
KNOC	Korean National Oil Corporation
LNG	liquefied natural gas
METI	Ministry of Economy, Trade and Industry
ONA	Office of Net Assessment (US)

OPEC	Organization of Petroleum Exporting Countries
OSCE	Organization for Security and Cooperation in Europe
PCA	partnership and cooperation agreement
SCO	Shanghai Cooperation Organisation
TEN	Trans-European Networks
UNEP	United Nations Environment Programme
WEO	World Environmental Organization
WHO	World Health Organization
WSSD	World Summit on Sustainable Development
WTO	World Trade Organization

Introduction

What Is Energy Security?

What is energy security? The European Commission and the International Energy Agency define it as the provision of reasonably priced, reliable and environmentally friendly energy. This would certainly be an acceptable general definition. However, as usual, the devil is in the details. What does reliable energy supply actually mean? What price, both financially and politically, are we prepared to pay for it? And how can we make sure that not only the interests of environmental protection, but also of transparency, human rights and democracy are not forgotten when energy resources are exploited? This is what the current political debate is all about: which way to more energy security is the correct one?

The energy question also illustrates how foreign policy has changed in the 21st century. In the search for oil, money and power, there seems to be no more room for noble principles of international law and the subtle instruments of international diplomacy. The gloves are off in the battle for the last resources; securing national energy supply has become the tough realpolitik for every country. Alliances are formed not with those that we like, but with those that we need. This book describes how the balance of power in the world is shifting, already today, according to who controls the remaining energy resources and access to the global market. It reveals how politicians must respond to solve the looming resource conflicts cooperatively and peacefully.

Russia, China, the European Union (EU) and naturally the US, are the four main players up against each other in the Great Game of the 21st century. Unlike the 'Great Game' of the 19th century, when Russia and England competed with each other over the control of Central Asia, today's game is not only about wrestling for the political and economic influence zones, it is also about defining the rules of the game for the energy markets, in particular, and for the world of tomorrow in general. There are two opposing philosophies: a new great politics of power, as the US has undertaken thorough military and political restructuring of the Middle East, and as Russia and China have pursued expansive policies through their state-owned energy companies in Africa and Central Asia; or a policy that – in order to ease resource conflicts – focuses on environmental protection, saving of energy, renewable energies and international cooperation.

Russia plays a key role in the future of energy supply for Europe and Asia. Therefore, it is important to understand where Russia is heading and who is in control in the Kremlin. The development in Russia's domestic policy also influences its conduct towards its neighbours and the rest of the world. As during the

Soviet Era, autocratic domestic policy and imperial foreign policy always go hand in hand. This time, the new great power politics of Russia focuses on the power of Gazprom, not on the weapons of the Red Army. The European Union must decide how it intends to deal with the uncanny colossus to the East. Will Russia become a strategic partner, as is always stated in the pleasant-sounding declarations of European summits, or is there a competitor and adversary growing in the East, against whom Europe must prepare itself politically and economically?

Energy and foreign politics are also closely linked for China and India, the two upcoming powers of the 21st century. The region and its dynamic economic growth are highly dependent on energy imports, both from the Middle East and from Russia and Central Asia. Asian emerging markets and their economies therefore react very sensitively to crises and wars in those regions.

The majority of the oil and gas reserves in the region itself are found below sea level. Sovereignty over these regions is fiercely fought for, causing an increase in tension between China and its neighbours and hampering regional cooperation in other areas. In addition, China's state-owned petroleum companies are pushing very aggressively on the global market, without any regard for environmental problems and human rights.

The EU is the greatest economic power in the world; however, it does not pursue a comprehensive energy policy. But without a common European energy policy, there can be no convincing EU foreign policy. Otherwise, the EU will remain prone to being blackmailed in a central economic question of survival. As Europe will be dependent on energy exports from its southern and eastern neighbours for the foreseeable future, the common energy policy must not be restricted to the EU in the narrower sense, but must include the neighbouring countries in Eastern Europe and the Near and Middle East. The EU's relationship with Russia is of major importance in this regard. As Russia connects the double continent Eurasia geographically, energy transport and trade not only link Eurasia economically, but also and increasingly, politically. Other key countries in this context are Ukraine, Turkey and the states of the Southern Caucasus. They serve as a political bridge and as transit countries for the energy imports into the EU. Therefore, these countries must be brought closer to the EU politically and economically.

However, Europe's energy policy must also develop alternatives to the growing dependence on energy imports from its neighbouring regions; for example, by further improving the energy efficiency of the European economy and by extending the old continent's leading role in renewable energies.

In the search for the last oil and gas reserves on the planet, both the environment and democracy are under pressure. Exploitation of natural resources is threatening the once pristine environment in the Arctic, in the oceans, and in tropical and Nordic forests. Corridors penetrate mountains, ice and forests to facilitate the transport of oil and gas. Poisonous waste and accidents threaten the indigenous fauna and flora, thus depriving the local agriculture and hunting industry of their basis for existence. Usually, the local population is not consulted when the ministry of energy in the capital or the multinational oil company build and invest in the area.

Where oil rules, local democracy and ownership are normally neglected. Violation of human rights and the disturbance of the traditional way of life of the indigenous population lead to social conflicts and political instability. Often, lobbies of indigenous people and international environmental associations receive no governmental assistance in their attempts to stop the colonization of the last natural paradises by international energy companies.

When the era of fossil fuel nears its end, the question will come up – possibly sooner than expected – of what will come next. Not every drop of oil, not every cubic metre of gas and not all coal will be extracted. Natural challenges and political crises will render the exploitation of the remaining fossil resources difficult and expensive. The poorest of the world will therefore be the ones most affected. In future, they will simply be unable to pay their oil and gas bills. The energy industry's search for alternatives is therefore well under way. Two diametrically opposing strategies are pursued: increased extension of nuclear energy, possibly complemented by futuristic fusion reactors on the one hand, and a sophisticated environmental policy that focuses on energy saving and renewable energies on the other. However, in the era of international terrorism worries about nuclear proliferation are also growing. The future therefore belongs to renewable energies, not only from an environmental, but also from a security point of view.

Nevertheless, the move away from oil not only requires corresponding changes in the economic course of the individual energy companies, but also a macroeconomic decision from every state with regard to its energy mixture; in the end, it can only be achieved through international cooperation: through international energy diplomacy.

Humanity is facing the choice between a peaceful decision on its common energy future or wars for resources in the near future. The system of international contracts and institutions that provides rules for the area of international energy policy, however, is fragmentary. A new world environment organization and an agency for the promotion of renewable energies could replace established institutions, such as the International Energy Agency.

What would a worldwide energy policy look like, in which not the rule of the strongest, but the rule of law prevails? How can diminishing resources be managed jointly, and how can wars for resources be avoided? How can the permanent destruction of pristine natural regions to access the last natural resources be prevented? And what will an international energy policy that not only provides secure, reasonably priced energy, but also answers humanity's demand for environmental protection, eventually look like?

Chapter 1

Facing a New Energy Crisis

The world is facing a new energy crisis. Unlike the oil crises of the 1970s and 1980s, this is not one of the known cyclical price increases, but rather a long-term trend towards scarcity of resources, caused by the entry of important emerging economies such as India and China onto the global energy market. Reserves of oil and gas are diminishing while demand is increasing. Only coal reserves still appear to be sufficient for a while. However, the combustion of all coal reserves would place an enormous burden on the environment, as worldwide emissions of the greenhouse gas carbon dioxide would increase even further. Apart from the energy crisis, politicians and the public are increasingly aware of the imminent climate crisis. After all, the first effects of global climate change can be observed today in the Arctic and other sensitive ecological systems.

Furthermore, the looming energy crisis threatens to shift the political balance of the world. The remaining oil and gas reserves are concentrated in the Persian Gulf, in central Asia and Russia. Europe therefore faces the risk of becoming dependent on politically unstable countries with undemocratic constitutions, the new great energy powers of the future. However, if Europe and other countries succeed in using renewable energies, the power balance in the geographic dynamics could once again shift in their favour.

Business as usual

The International Energy Agency (IEA) in Paris was founded after the first oil crisis in 1973. It monitors the development of worldwide energy markets. Its annual *World Energy Outlook* publishes regularly collected data on trends in energy consumption, its exploitation and price development in all major industrial and emerging countries. It is worth having a look at the latest figures published by the IEA to understand what dramatic developments are looming on the global energy markets and the political challenges associated with these.

In *World Energy Outlook 2005,* the IEA has drafted a scenario for the period from 2005 to 2030. The basic assumption is that the energy policy of the major industrial and emerging countries will not significantly change during this period. Americans would call this 'business as usual'. If today's trends continue, based on the expected development of the global economy, the experts at IEA have worked out that worldwide energy consumption will increase by 50 per cent by 2030.

In accordance with this trend, energy supply in 2030 will mostly still rely on fossil fuel resources as well. Global consumption of oil, gas and coal will accordingly increase further. The bulk of the increase will be experienced in the major emerging markets of India and China. As a consequence, emissions of the greenhouse gas CO_2 will increase by 1.6 per cent per year. The goal of the UN climate convention passed at the Rio Summit in 1992 was to reverse this trend by the year 2000 and to reduce CO_2 emissions to below the level of 1990. Achieving this goal is no longer possible today. If current trends continue, the nuclear energy percentage will decrease as fewer new reactors are planned than are shut down. The percentage of renewable energies, such as sun, wind, water and biomass will increase faster than that of all other energy forms. However, as the growth of renewable energy technologies starts from a low level, they would only cover 2 per cent of the entire primary energy demand, even in 2030, despite the expected high annual increase of 6.2 per cent.

The annual energy report of the German Federal Ministry of the Environment confirms the IEA figures. According to the report and if current trends continue (that is if renewable energies moderately increase while nuclear energy is abandoned), the calculated total percentage of non-fossil fuel, that is nuclear energy and renewable energies, will decrease rather than increase. The percentage of oil and natural gas in the energy mix would increase from 59 to 70 per cent in the same period. In addition, the German Federal Ministry for the Economy points out that European production from oil and gas fields in the North Sea will diminish further, meaning that dependence on imports from Russia and the Middle East will increase.

In this context, the ministry experts are worried by the political development in both regions. Political transformation processes in Russia are far from finalized. Nobody knows whether Russia will develop a stable democracy and market economy in the long run. From a security point of view, the Middle East is crisis region number one. Consequently, reliable energy supply from this region is not guaranteed. Besides, neither Russia nor the majority of the Gulf states are members of the World Trade Organization (WTO). This means that, should there be any trade conflict, the mediation instruments of the WTO do not even apply to them. The German Federal Ministry for the Economy therefore predicts that Germany's energy security will probably decrease as it becomes increasingly dependent on the two export regions mentioned.

All in all, the IEA expects that existing fossil fuel resources will still suffice to supply the global economy in 2030. However, the forecast of the IEA contains some uncertainties. Its basic assumption foresees the discovery of new oilfields in the next few years and thus a further increase of refinery capacities, and consequently it expects only a moderate increase in the global oil price – and of the coupled gas price. In 2005, it predicted an average oil price of US$35 per barrel for 2010, increasing to $37 by 2020, and to $39 by 2030. However, in mid-2006 during the political crises in Iraq, Iran and Lebanon, the price per barrel reached almost $80, thus more than double the predicted price. In early 2008, the oil price crossed the symbolic threshold of $100 per barrel for the first time. And there is no end in sight.

The IEA experts rightly point out that natural catastrophes, political crises and wars may upset energy security for the consumer countries, even if quantities of global oil, gas and coal reserves are sufficient. However, the IEA is particularly worried about the growing asymmetry between few exporting and many consumer countries. If current trends continue, the world's dependence on a few oil and gas-exporting countries, in particular, from the Near and Middle East, will grow dramatically. At the same time, the dependence of Western Europe on natural gas imports from Russia will increase. Unlike North America, the third pole of the industrialized world, Western Europe and East Asia have themselves only minor fossil fuel reserves. Apart from the loss of energy security, the IEA experts are first and foremost worried about increasing emissions of greenhouse gases if fossil fuel combustion is not reduced. The conclusions of *World Energy Outlook 2005* therefore read: 'It is widely recognized that the outcome of this reference scenario is undesirable and non-sustainable.'

Peak oil – the end of cheap petroleum

A ghost haunts the energy industry. Its name is 'peak oil' and it prophesizes the end of the oil era – or at least the end of the availability of a cheap, seemingly inexhaustible lubricant for the global economy. In a recent newspaper advertisement, American oil company Chevron said: 'It took us 125 years to use the first trillion (1000 billion) barrels of oil. We'll use the next trillion in 30.' They refer to the production of the entire industry, not only that of the Chevron company. The question is whether the next trillion will also be the last trillion. A new branch of science deals with the time and the consequences of peak oil. Kjell Aleklett, professor in physics at the University of Uppsala, Sweden, has founded the Association for the Study of Peak Oil and Gas (ASPO) and scientifically researches this subject. Aleklett publishes news and newspaper articles that prove that oil and gas are taking a turn for the worse. In his blog (www.peakoil.net), a discussion forum on the Internet, geologists, energy traders and oil exploration experts discuss their common worry that the end of the oil era could be closer than we think.

Peak oil defines the time when worldwide total exploration of petroleum has reached its peak and from then on it will slowly decline. According to data from oil company BP, the estimated crude oil reserves will last for another 40–50 years. Since the beginning of the 1980s, more oil is being extracted than new oil is found, and the gap is widening further. Long-term investment in many old oilfields no longer pays. The old ramshackle production systems and the rusty tankers are therefore used until the well has run dry. After that, as the oil era is in any case coming to an end, they will be scrapped.

The time that the reserves of the major oil producing countries will last varies. While Saudi Arabia produces 1.5 per cent of its reserves annually, this figure amounts to 3 per cent for Africa and 5 for Russia. Roughly calculated, Russia's oil reserves will thus be used up within 20 years. In addition, there are other resources that cannot at present be exploited at an economically and technically reasonable cost. However, this may change if the oil price goes up. After all, unconventional oil sources, such as oil slate and oil sand, also exist, and these would last considerably longer.

The usually cited ratio of reserves to statistical consumption, however, is misleading, as it incorrectly suggests that constant production could be maintained until all reserves are exhausted. Therefore, what is important is the time from which production will start to decline. ASPO believes that this point could be reached as early as 2010. The German Federal Institute for Geosciences predicts peak oil by the year 2025. Then, at the latest, oil will no longer be available as a cheap and seemingly inexhaustible natural resource, and the economy of scarcity – or the search for alternatives – will begin. The most important consequence of the lack of oil after peak oil is increasing prices, as supply can no longer satisfy demand. Not everyone will be able to pay these prices. They will experience a new form of poverty: energy poverty.

As mentioned above, the remaining reserves are concentrated in fewer and fewer regions. Approximately 70 per cent of conventional crude oil and 65 per cent of natural gas reserves can be found in a relatively confined region. This region, the so-called strategic ellipsis, spans from the Middle East via the Caspian region to north-western Siberia. As a result, this is the area that foreign policy strategists in Washington, Moscow and Beijing focus on. Here, the Great Game for the last resources of the fossil era will take place. Other regions, first and foremost Europe, with little of its own energy, become increasingly dependent on imports, unless they succeed in replacing fossil fuels by renewable energies and drastically increasing the efficiency of their energy utilization.

North Sea oil is an illustrative example of what the end of the oil era could look like. Following the first oil price shock in the beginning of the 1970s, the discovery of new oil reserves beneath the North Sea helped Western Europeans become less dependent on imports from the Arabian OPEC countries. Even if oil from the North Sea was considerably more expensive than the imports from the Persian Gulf, the monopoly of OPEC was broken. New oil reserves were also discovered in Alaska. Russia and the Central Asian states entered the international oil market in the 1990s.

Today, North Sea oil production is declining. Most oil wells will be depleted by 2020. Already today, the UK has once again become a net importer of petroleum products and natural gas. Norway, the second major oil power in the North Sea, is developing gas fields in its Arctic North and plans a pipeline along its coast as competition for the Baltic Sea pipeline and Russian trade with liquid gas. For the foreseeable future, Norway will be Western Europe's only energy exporter.

Between 1995 and 2005, the average cost of producing one barrel of crude oil increased from US$5 to US$10. This is because the majority of the oil wells close to the surface and thus cheap to exploit are slowly running dry and the exploitation of less accessible regions is more expensive. The cost of production equipment, such as steel or oilrigs, is also increasing. Already, a serious lack of skilled labour is becoming noticeable for the increasingly complicated and technically more demanding oil production in the tropics, underneath the seabed (offshore) or in the Arctic. In countries with poor education systems, such as the oil-exporting states of Central and Southern Africa, this lack of skills is reaching dramatic dimensions. Western technicians and experts are reluctant to work in these countries because of the bad security situation. For instance, the staff of international oil companies are regularly attacked or abducted in Nigeria. The consequences of a lack of technicians are production losses and an increasing number of accidents.

Politicians have also discovered that the oil era is coming to an end. In his State of the Union Address at the beginning of 2006, US President Bush surprisingly demanded that America should be freed from its dependence on oil. Bush even admitted that the US was 'addicted to oil'. The American Department of Defense has engaged the pope of alternative energies, Amory Lovins, to advise them on how to win the oil endgame. For the first time, the Swedish government has submitted a plan on how an advanced industrial country can manage without oil imports by the year 2020. In an advertisement series, energy company BP no longer calls itself 'British Petroleum', but 'Beyond Petroleum' – thus indicating that it sees a perspective for its business even beyond petroleum. Oil is still the main business of the BP group.

The 'end of oil' also has a climate policy dimension. Sweden's neighbour Norway has announced the offsetting of the carbon emissions of its oil exports by protecting rainforests in developing countries such as Brazil and Indonesia.

Nevertheless, investment is still being made in the continuation of the oil and gas era. The IEA expects that until 2030, a total of US$13 trillion will be spent worldwide on the future of energy supply. However, a decision has to be taken in which technologies those investments should be made. One of the major oil companies in the world, British BP, plans annual investment amounting to US$15 billion, as well as another 2 billion through its Russian subsidiary TNK-BP. This high investment requirement is also the multinational energy companies' justification for their skimming of profits in times of high oil and gas prices.

Can peak oil be delayed? Can the oil era be extended? If one studies the full-page advertisements that companies such as Chevron and Shell have published on the subject of peak oil, the answer is 'Yes'. However, the question is: at what price?

Oil sands – Canada's Saudi Arabia

The saga of the allegedly inexhaustible potential of the oil sands of Canada and Venezuela that has been preached for decades is illustrative. In the early 1970s during the first oil crisis, the German magazine, *Stern*, published an article on the Canadian oil sands with spectacular photographs. There, in the west Canadian province of Alberta, journalists found the future of our energy supply beyond OPEC and the oil embargo.

Oil or tar sands contain potentially gigantic reserves of still unexploited crude oil reserves. These are layers of earth, slate or sand saturated with oil that can be found in numerous countries. The largest deposits lie in the Canadian province of Alberta, in the Orinoco river basin in Venezuela and in the vastness of Russia. The Canadian reserves alone are estimated to be larger than the conventional oil reserves of Saudi Arabia. With the help of these great oil sands, Canada could thus, theoretically, become a competitor for Saudi Arabia on the global petroleum market and significantly change the geostrategic balance.

Obviously, this would have happened long ago if it were not for a few problems. Not only the cost, but also the expenditure of energy, water and natural resources are significantly higher compared to conventional energy production.

Oil sands are exploited with shovel excavators in open-cast mines. Those who have seen the German lignite surface mining regions of Garzweiler or the Lausitz know that an unromantic moon landscape is left behind afterwards. Any subsequent renaturalization measures, if implemented at all, cannot restore the lost natural features but rather create a replacement landscape with few species and little variation. Pristine land, as still existing in parts of Canada, for example, is impossible to recover. Besides, oil sand mining in Canada's west is by some dimensions greater than surface coal mining as we know it in Germany. Entire forests are cleared for this today, rivers are diverted and people resettled. If accidents happen, rivers, lakes and drinking water could be contaminated with oil.

One quarter of the energy content of the produced crude oil must be spent on the expensive exploitation and preparation of the tar sands as well as subsequent renaturalization. The petroleum is extracted from the sand using water steam heated by natural gas combustion. Plans exist to build a new transcontinental gas pipeline from the Arctic through the Mackenzie valley to support the new oil sand development. The Canadian government is even considering building a special nuclear power station to provide the thermal energy necessary for water steam production. Apart from the high energy consumption, the required quantities of water also pose a problem. The water must be purified afterwards in an expensive process and must be returned to the denaturalized river system via specially built waterways.

Production of marketable oil from slate and sand in the remote exploitation areas of Siberia or the tropical Orinoco would certainly not be cheaper. The environmental impact of large-scale exploitation of oil sand in open-cast mining there can only be estimated. However, while the media and environmental groups in the democratic system of Canada ensure a modicum of transparency and control, oil production on the outer borders of the tropics or the Arctic takes place largely with the exclusion of the public.

The oil sands of Canada are only one – be it a very illustrative – example of the fact that the exploitation of new oil and gas reserves is becoming increasingly more expensive, more difficult and also increasingly more dangerous for man and the environment. The first oil wells in North America and in the Middle East were discovered by virtue of the oil emerging naturally on the surface. Many of these wells had been known since ancient times: In Mesopotamia, tar and oil were used to seal ships and as a miracle medical ointment. In what is today Azerbaijan, the fire cult of Zarathustra originated at a place where oil wells emerging on the surface of the earth self-ignited. Such easily accessible wells are no longer found today, and the majority of those known have been depleted. Drilling must go deeper, through harder stone, through ice or beneath the seabed. As the global petroleum industry is now entering the last wildernesses in the search for the last hideouts of Black Gold, the price nature must pay also increases.

Natural gas – an alternative?

At the same time, it has become evident that the long-touted alternative, natural gas, is not available either in limitless and cheap quantities. 'Peak gas', that is the peak of worldwide gas production, however, is further away than the peak of oil production. Many countries have only just begun converting power plants and heat generation from oil or coal to natural gas. Natural gas offers a number of advantages. Only minor amounts of pollutants are generated during its combustion and its CO_2 content per energy unit is lower than for coal and oil. Many environmental politicians who advocate changing completely to renewable energies in the long run accept natural gas as an interim solution. However, two major problems that this environmental strategy faces are the gas price and the dependence on a few gas-producing countries, both of which have increased significantly lately.

Often, natural gas is present in already exploited oilfields, but has so far not been utilized sufficiently. In many places, the escaping natural gas is still being burnt and is not used for generating energy. What it does generate, however, is the greenhouse gas CO_2. Consequently, the potential of natural gas utilization from existing sources is significant. Besides, since systematic exploration started, several sizeable new natural gas fields have been discovered every year.

Natural gas is normally transported to the customer via pipelines. As it becomes more expensive to maintain the gas pressure in these pipelines over long distances, the rule of thumb is that gas pipelines should have a maximum length of 4000km. This is the reason why no global gas market exists, and only regional networks are in place so far.

This could change with the increasing trend towards the use of liquefied natural gas (LNG). Natural gas can be liquefied under pressure and at low temperatures. One advantage of this method is that it reduces the volume of the natural gas. Also, LNG can be transported by tankers to its destination, the same as oil. The largest liquefied natural gas plant in the world is currently being erected for the East Asian market on the Russian island of Sakhalin in the Pacific. It is planned to export LNG from the Arctic peninsula of Yamal to North America by tanker ships. In addition to Europe and the states of East Asia, Russia would thus have a third customer market for its gas exports. Nigeria and Algeria have also put their money into liquefied natural gas exports to Europe. Japan and China are particularly interested in pipeline-independent access to clean natural gas.

Thanks to the unlimited transportability of LNG by tanker, the natural gas market will become a global market. For this reason, LNG terminals are currently being constructed extensively along the coasts of North America and East Asia. There is a boom in shipyard orders for new tanker ships.

Until now, the market has not determined the price for natural gas. Most supply contracts provide for a long-term fixed price. Only in this manner was immense investment in the gigantic pipeline network possible, which, for example, connects Eastern and Western Europe. Russia's close alliance partners, such as the dictatorship of Belarus, have, until recently, received price discounts. In 2007, a new price deal between Russia's Gazprom and the government in Minsk was negotiated that transfers control of the Belarusian gas pipeline network to the Russian company. In exchange, the prices are still rising but not quite to world market levels. Even though the gas price for Ukraine is to be increased gradually, it is still kept at a low level by adding cheaper natural gas from Turkmenistan, Kazakhstan and Uzbekistan. Otherwise, the gas price traded on the global spot markets is pegged to the oil price. In future, once the two products are produced and traded independently of each other, this pegging will eventually be discontinued.

Natural gas is probably the natural resource most sensitive to political crises that is traded on the global market. Permanent pipelines are expensive, require long construction times and, afterwards, can no longer be moved or diverted as an oil tanker might be. Whoever relies on gas therefore relies on two things: the reciprocal dependence of producer and consumer, and the diversification of the sources. Europe is fortunate to be surrounded by natural gas-exporting countries. Even though some Eastern European countries purchase their gas almost exclusively from Russia, the EU as a whole has a diversified supplier structure. Besides, the EU tries not only to regard the countries that surround it as importers of natural resources, but to integrate them gradually into the common European market. The basic idea behind this is that mutual dependence will lead to political cooperation and eventually stability. The foreign policy concept of the European Neighbourhood Policy, which the EU also applies to structuring its financial aid to the countries of Eastern Europe and North Africa, therefore places the focus on joint management of the energy resources.

Coal

The energy resources that will last the longest are the worldwide hard coal reserves. Apart from Russia, the US, Australia and South Africa, China and India also have immense national hard coal deposits that are still far from being extracted. The IEA estimates that coal consumption in the latter two countries alone will increase by 60 per cent by 2030. Also in the US and Russia, coal is by far the richest fossil fuel source. Considerable underground beds of hard coal and lignite exist even in many European countries such as Germany and Poland, which otherwise have hardly any own fossil fuel reserves. However, if all known hard coal were combusted using current methods in thermal power plants and without a CO_2 filter, this would have unpredictable negative consequences for the global climate. Since neither the major emerging markets nor Russia and the US are going to give up using this cheap local energy source, technologies as clean and efficient as possible must be used for coal combustion to restrict the environmental impact.

In principle, three options are available to make coal combustion more environmentally friendly. Apart from highly efficient thermal power plants, coal can be liquefied and used as fuel with versatile applications. Also, it is possible to separate out the carbon dioxide that is generated during combustion of fossil fuel and to store this underground to prevent its release into the atmosphere. This technology is called 'carbon capture and storage (CCS)'. The European Commission announced in spring 2008 that it would support the construction of ten CCS demonstration plants in the EU to test the technical and economic feasibility of this emerging technology.

Increasing the degree of efficiency of conventional coal power plants is a first important step. Especially in countries such as Russia, Eastern Europe, India and China, all of which are among the greatest coal consumers in the world, there is an enormous potential for improving energy efficiency. However, the increase of energy efficiency per unit of coal has its technical limits. The maximum achievable degree of efficiency is 65–70 per cent. During subsequent flue gas cleaning, that is filtering of the sulphur and other toxic substances from the chimneys, part of the generated energy is lost. Unfortunately, the only way to build modern power plants with expensive exhaust gas purification economically is as large-scale plants. Consequently, they are not well suited to be used as combined heat and power plants for industry and households. This is because heat can only be transported economically over relatively short distances. Most waste heat from large power plants is therefore radiated into the atmosphere.

Coal liquefaction has a bad reputation, stemming from its history during World War II. During the Third Reich and later in the German Democratic Republic (GDR), fuel replacement was produced from lignite in the 'chemical triangle' in Saxony. In the meantime, however, the quality of mineral oil products made from coal has improved considerably by the use of catalytic converters, information technology and modern measuring systems. Harmful substances can be separated without difficulty in the liquid phase. In test plants, the degree of efficiency is up to 95 per cent. Moreover, coal liquefaction solves one problem associated with many renewable energy technologies: the resultant liquid energy source is easy to store and can also be used as vehicle fuel. If oil prices remain high, coal liquefaction is also sustainable from an economic point of view. It is therefore a real alternative to petroleum. The South African company Sasol, which developed its coal liquefaction method amid the economic boycott that prevailed during the Apartheid years, produces liquefied coal at US$25 per barrel, significantly below the price for crude oil expected in future. Even for German coal, production costs of approximately US$60 per barrel are considered realistic. In 2006, the world oil price was above this value for months. Coal deposits can be found at so many places and in such quantities that the worldwide coal market is less sensitive to political crises than the oil and gas markets. Today, most of the coal imported by Germany comes from Australia, a stable democracy. However, from a climate point of view, liquefaction of coal is even more unfavourable than its combustion. This is because additional energy must be used for converting coal into fuel.

In principle, it is possible to make coal combustion more climate-friendly by filtering out the greenhouse gas CO_2. In this case, the significance of coal combustion could increase once again even in Western industrialized countries. The same CO_2 separation and storage technology could be used in oil and gas power plants. Electricity company and Swedish energy group Vattenfall is planning the first CO_2-free power plant operating with lignite in Lausitz in the east of Germany. British Petroleum is in the process of developing similar projects for California and Scotland. However, on the way to a routine use of this technology, a few problems still need to be solved. When current CO_2-filtering technologies are used, the energy efficiency of the power plants decreases significantly. The technology thus becomes too expensive, above all for developing countries. However, the cost difference could be met from an international climate protection fund. Compared to the thousands of traditional coal power plants that are currently being built primarily in China, India and other developing countries, the current Vattenfall pilot plants in the Lausitz region, as well as similar plans that German electricity company RWE have for the state of North Rhine-Westphalia, are nothing more than the proverbial drop in the ocean. The decisive question for the future of coal is: will there be no more than a few token CO_2-free pilot plants, or will the costly modern separation technology be used in modern power plants everywhere in the world?

However, before a decision on the large-scale use of this new technology is made, the basic problem of where to store the separated CO_2 afterwards should be solved. So far, proposals provide for either putting it in old oil and gas storage deposits, in salt deposits or beneath the seabed. As yet scientists know too little about the long-term behaviour of these CO_2 bubbles and are not in a position to recommend large-scale use of this technology with a clear conscience.

Energy poverty

The new curse affecting the poorest countries of the world is called energy poverty. Not only the wealthy industrialized countries or the growing economies China and India, but also the poorest of the poor in Africa, Latin America or Central Asia, are among the 85 per cent of all countries that must import oil. Increasing prices have a disproportionately high impact on these countries. If energy prices rise further, the poorest developing countries are threatened by a new debt crisis. The World Bank has worked out that an increase in crude oil prices of US$10 costs the industrialized states half a per cent in economic growth. For the poorest countries, where the percentage of energy costs in the manufacture of goods is usually significantly higher, the losses may be up to three times as high.

Energy poverty does not only hit the economy of these countries, but also their citizens. Due to increasing gas prices, tenants in Ukraine will soon no longer be able to pay their heating bills. Those who cannot afford the expensive energy will have to freeze. Schoolchildren in Afghanistan are no longer able to study in the evening because electricity is cut off. The chances to advance in life are fewer for people living in countries, suburbs or households with little energy. Many governments of the poorest countries must implement savings in the education and health system to settle the oil import accounts. Energy poverty blocks the way to development.

Some countries that are exporters of energy on the global market have citizens who suffer from energy poverty. This is due to regional and social inequalities. In Nigeria, the major exporter of crude oil in Africa, long queues of cars form at filling stations in the capital Lagos. Even though Nigeria exports oil, it lacks refinery capacity. The Niger delta, where most of the oil in the country is produced, is at the same time one of the poorest regions of Nigeria. Most villages are not connected to the electricity grid. Wood is used for fuel.

Even in Russia, the largest producer of natural gas worldwide, numerous villages are cut off from the modern era. In winter, supplies are confined to the towns. In some villages of Siberia, people can see the flares of natural gas drilling rigs in the distance but they have to use wood for fuel.

Energy poverty does not only affect the economy of poor countries, resulting in a loss of opportunities for the people living there, it also disturbs the stability of fragile states and young democracies. Unilateral development of the export-oriented energy sector takes place at the cost of other sectors of the economy. Existing social differences grow and political tensions increase. In this context, it normally makes no difference whether private multinational groups or state-owned energy companies control the industry. In Russia, Iran and Venezuela, nationalization of the energy industry was accompanied by a loss of democracy and an aggressive foreign policy.

The best chance that poor countries have to escape the energy poverty trap is to go their own way towards sustainable energy supply. Today, the inefficient economy of the least developed countries consumes more than double the energy per unit of economic output than that of Western industrialized countries. The energy-saving potential would be enormous were investments made in the corresponding technical equipment. Another alternative to the import of expensive fossil fuel would be the more effective use of local resources. In many agriculturally structured developing countries, biomass can be used for heating and for electricity generation, and ethanol from grain could be used as a petrol replacement. Wind and solar energy are good alternatives, in particular in remote locations that the national electricity network cannot reach. Modern development cooperation therefore focuses on the access of all to affordable energy from renewable sources.

Climate security

The new energy crisis is simultaneously also a climate crisis. If we talk about energy security, we must also talk about climate security in the future.

The new report of the Intergovernmental Panel on Climate Change (IPCC), the council of scientists that observes worldwide climate change for the United Nations, was published at the beginning of 2007. The scientific findings are alarming. Already we know that the concentration of gases that influence the greenhouse effect (carbon dioxide (CO_2), methane and nitrogen) is higher today than at any other time in the past 650,000 years. Between 1999 and 2004, the CO_2 content of the atmosphere increased by 0.5 per cent per year. Overall, the temperature of the Earth's surface has increased by 0.65° C since the beginning of the 20th century.

In his new work, *Collapse*, Jared Diamond, professor of geography at the University of California and winner of the Pulitzer Prize for his book, *Guns, Germs and Steel* on the natural bases of different human civilizations, deals with the question of how different cultures respond successfully to ecological crises or collapse.

A few summers ago, Diamond visited the two dairy farms of Huls and Gardar and found amazing similarities. Both were among the largest and technically best-equipped farms in the region, and their owners were pillars of their communities. The problems of the two farms were also alike: both were situated too far north for cattle to graze on the pastures throughout the year, and they therefore operated on the border of profitability for a long time. The major difference between Huls and Gardar, however, consists in their present situation. The Huls farm in the state of Montana is a prosperous family business in one of the districts with the highest population growth in the US. The Viking farm of Gardar, on the other hand, has been lying destroyed and abandoned on the west coast of Greenland for 600 years.

Based on these and other cases, Diamond investigates how the destruction of the natural means of existence has led to the degradation and the complete disappearance of historical civilizations. In comparison, he describes societies that have handled the same or similar challenges more successfully, and looks for parallels in modern societies. His basic question is why societies undermine their own means of existence by economic and cultural decisions, even if the signs on the wall are clearly legible. At the end of the book, he proposes that we should learn from the historical failure of others.

The civilizations of Easter Island or of the Maya were also victims of historic destruction of the environment. The culture of the Vikings survived on the Faroe and Shetland Islands and in Iceland, but did not adapt to the more extreme conditions in Greenland. Other cultures responded to initial environmental destruction with social innovations, for example, the introduction of sustainable forestry in Japan and Germany in the 19th century.

The end of the Arctic

Greenland and the fall of the Viking people stand in the centre of Jared Diamond's book. However, today's Greenland is also threatened by environmental changes. On 8 November 2004, a few days after US President George W. Bush was re-elected, the *Arctic Climate Impact Assessment* written by a team of 300 internationally renowned scientists was published. The initiative had come from the Arctic Council, the interstate organization of Arctic bordering states. The US, a member of the Council, had blocked the publication of the report before the presidential elections.

The most important finding is that the Arctic is currently heating up at twice the rate of the rest of the Earth. The ice cap over the Arctic Ocean and the ice shield on Greenland are starting to melt. Should the entire inland ice of Greenland melt, the worldwide sea level would rise by 7m. The increase would be even more dramatic if the ice of the Antarctic were to melt as well. The destruction of Western Antarctic ice shield, observed by environmental satellites over the past few years, could consequently be a greater strategic threat for worldwide security than the Iranian nuclear programme.

The Arctic Council is an international organization of a special character. Apart from the bordering states, six world umbrella groups of indigenous people are also among its members. To take Russia as an example, the Arctic Council view on the region and its natural resources differs from the central government in Moscow. The Arctic is on the political and economic periphery for almost all member states of the organization. Thus, the indigenous organizations are the driving force of the organization.

The International Arctic Science Committee (IASC) was created from the merger of 18 national scientific academies. Scientific academies play an important role as independent advisers to their governments – above all in Russia and the US. The US Academy has warned its government for years about the consequences of global climate change. Already during the Soviet era, the Russian Academy of Sciences was a place of refuge for ecologically minded researchers and intellectuals.

Within the Arctic Council, it is primarily the Canadian government that has made climate change in the region a topic of public diplomacy. In this context, the government in Ottawa works closely with the secretariat of the Council in Copenhagen, indigenous representatives and non-governmental organizations. As part of a media campaign, the Arctic Council has also gained the support of the *New York Times*, which has reported continuously on the problem for the past 18 months. Since then, other international media that use the *New York Times* as inspiration have since taken on the subject.

For the indigenous cultures of the Arctic, the changing climate accelerates the need for necessary adaptations. The economy and culture of the Inuit, for example, are closely linked to the use of only a few animal species. Due to the decline of marine mammals and the fishing yield, local diet must be changed to include imported processed foodstuff. This is accompanied by an increase in modern ailments linked with refined diets. Tools and artistic craftwork made from animal products represent the second pillar of the Inuit economy. This basis of what has been a self-sustaining economy is also threatened, without any alternatives in sight.

In addition, the foundations of modern economies are threatened in the changing Arctic. In the Russian Federation, industrialization – and thus the replacement of indigenous economies – is significantly further advanced since the Soviet era, than is the case in Canada, Alaska or the Scandinavian states. Oil and gas are produced in Arctic Siberia, while large-scale forestry is undertaken in the tundra. Due to climate change, this exploitation is also facing new challenges. Industrial plants, residential buildings, streets, airports and pipelines were all built on permafrost. If this soil, which would normally be permanently frozen, starts melting due to higher summer temperatures, foundations and roadways will be damaged. Today, damage to buildings can already be seen in all Arctic settlements throughout Russia. Newer industrial plants have foundations that go through the permafrost. However, these can only be built at enormous cost.

Containing the climate catastrophe

A rise of 1m in sea level, which we might face as early as the end of this century, can no longer be managed by normal coastal protection measures such as dyke construction. Flat coastal regions and small island states, such as the Maldives, Tuvalu or Tonga, would be completely or partially flooded and disappear. This would lead to regional conflicts over the diminishing areas of arable and cultivable land, as well as to streams of refugees from the affected regions.

The Office of Net Assessment (ONA) is the internal think tank of the US Ministry of Defense – better known as the Pentagon. ONA's task is to reflect on long-term dangers that might threaten the US and its security. At the end of 2003, when the top management of the Pentagon was busy executing the war in Iraq and the fight against terrorism, ONA published a report on the climate catastrophe. The Pentagon masterminds fear a horror scenario combination of natural catastrophes, streams of refugees and wars for resources, which might cause the death of millions of people over the next 20 years. How can the US prepare itself against this new danger? One option would be through preventative politics, to reduce the risk of permanent climate damage and its security consequences. Another possibility would be to erect a protective barrier around their own country – not only against increasing water masses, but also against refugees from those parts of the world that cannot afford expensive adaptation strategies and resettlement measures for the affected populations. The question is whether we wish to live in such a world. Maintaining a protective barrier against the environmental refugees of the future would probably only be possible through an internal dictatorship and external military defence.

The creators of Hollywood's fantasy worlds also dealt with such horror scenarios for a while. In his film *The Day After Tomorrow*, producer Roland Emmerich tells the story of a climate researcher who – along with the rest of the world – is surprised by the start of abrupt climate change. Gigantic storms happen, Los Angeles is devastated by tornadoes, New York sinks beneath an enormous tidal wave, while the entire northern hemisphere experiences an ice age. And while trying to escape, the vice president of the US dies. His features resemble those of Dick Cheney, Vice President to George W. Bush. Cheney, himself a former top manager of the oil service company Halliburton, is one of the architects of the current US energy policy. His energy plan focuses on increasing supply, above all by exploiting additional deposits of fossil fuel, instead of conservation measures and climate protection.

The inhabitants of coastal regions and agricultural lands threatened by erosion and the inhabitants of especially sensitive ecological systems in the tropics and the Arctic are particularly affected by climate change. However, while indigenous people and the local fauna see their means of existence melt away, bordering states such as Russia hope to find new economic possibilities. Russia and the US bank on the waterways of the Arctic remaining open in future, even during winter, as a consequence of increasing temperatures. In such an event, oil and liquefied natural gas from the North American Arctic and from the Russian Arctic Ocean could flow unobstructed onto the global market transported by tankers. While individual resourceful companies could benefit from climate change, economic scientists – such as economist William Nordhaus from Yale – agree that the overall costs of climate changes to be expected will in all cases be considerably higher than the profit from any possible positive effects. Former chief economist of the World Bank Nicholas Stern calculated for the British government that the macroeconomic costs of climate change would amount to 5–20 per cent of global economic output. Not included in this calculation are the costs of wars and crises that are likely to occur because of attendant political distortions.

The collapse of a modern society in the context of the climate catastrophe could be observed after Hurricane Katrina destroyed the city of New Orleans and wide areas of the US Gulf Coast in August 2005. Immediately after the storm, American environmentalists posed the question as to how far Katrina was caused by global climate change. Although it is not possible to link individual natural catastrophes clearly to the long-term trend of global warming, there are clear statistical indications that the intensity of the storms in the subtropical weather system of the Caribbean has constantly risen over past decades. Climate researchers assume that the reason for this is the increased surface temperature of the seawater in the Gulf of Mexico causing instabilities in the lower atmosphere and thus faster movement of the air masses. The US National Hurricane Center alleges that the number and intensity of the storms vary as part of a natural cycle. However, this cycle seems to be overlapped by man-made climate change that causes peak activities to increase.

Finally, Katrina illustrates how expensive it will be to protect the coasts of the world against increasing floods and growing storms. Even if the US succeeds in securing the rebuilt New Orleans with dykes, like those used in the Netherlands, before the next Category 5 storm, a great number of coastal and island states lack the means for such costly infrastructure investments. The Bahamas and Bangladesh cannot surround themselves by dykes. Therefore, these countries have been clamouring for years at international climate negotiations for the creation of an investment fund to cover protective measures against climate change.

The double energy crisis

The world is currently heading towards a double energy crisis. The data of the International Energy Agency (IEA) and of the Intergovernmental Panel on Climate Change (IPCC) must be read and interpreted within that context. This means that the energy policy of the future cannot focus solely on the declining availability of fossil fuels but must also consider the restricted ability of natural ecological systems and of the atmosphere to withstand stress. If the reference scenario for the development of energy consumption of the next few decades is not sustainable, as the IEA states, we will need alternative development paths. There are three ways to react to the double crisis in energy and climate security.

First, we can reduce our energy consumption by the use of new technologies. Second, every country can reduce its import dependence by replacing fossil fuels, such as oil and gas, with renewable energy sources. Stronger focus on nuclear power, on the other hand, is not a good alternative, as it creates new dependence and increases the proliferation risk. Finally, binding rules for trade and investment, agreed at common institutions, such as the European Energy Charter and the World Trade Organization, could ensure a legal system for the energy policy. All these approaches only make sense if countries, at least within the European framework, act jointly.

Chapter 2

The Great Game for Measuring the World

The new problem child of the world community is called energy security. One could describe the political discipline aimed at making states more 'energy secure' as energy foreign policy. According to Germany's Minister of Foreign Affairs, Frank-Walter Steinmeier, maintaining global security in the 21st century will 'inseparably also be linked to energy security'. The dispute over worldwide energy resources has long since become an important element of the foreign and security politics of the major economic powers. It is part of a return to geopolitics, of the struggle for a new world order. Already today this struggle has resulted in foreign policy tensions and a shift in the global power balance. The US and Europe, and the upcoming economic powers China and India, compete against each other for secure access to the world's last oil and gas reserves. Russia's new imperial foreign policy is built on the power of its energy companies. The world is remeasured; the power balance between the major powers is redetermined. However, it is important in this context that legitimate competition for economic future opportunities takes place peacefully and follows generally accepted rules, and that other interests, such as the protection of the environment and human rights, are also taken into consideration.

Energy security in times of crisis

When Winston Churchill, then First Lord of the Admiralty, on the eve of World War I ordered the warships of the Royal Navy to be changed from coal firing to oil, his critics feared that the oil imported from Persia – unlike coal from Wales – would soon run out. When asked about the security of the new energy source, oil, in a time of crisis, Churchill had his answer ready: 'Safety and security lie in variety and variety alone.' England had this variety – unlike its soon-to-be enemy Germany – because of the oilfields that had just been discovered in Persia.

Churchill's principle that energy security essentially consists of ensuring supply from various sources is still applicable today. For its energy supply, no power in the world relies on the functioning of the global market alone. Energy security has always also been alliance – and foreign policy. Unlike in 1910, a central role is played today by the limits environmental and climate protection impose on the expansion of our global economy.

Furthermore, today the subject of energy security once again enjoys as much attention as it did in the oil crises of the 1970s. Since China, India and other major emerging markets entered the global economy, oil and natural gas are becoming scarce. Prices have risen to their highest level in history. Also, a number of political crises and natural catastrophes at the beginning of the new millennium have made it clear how vulnerable the worldwide energy system is. These include the war in Iraq, the crisis surrounding the Iranian nuclear programme, the new aggressive foreign policy of Russia, and also Hurricane Katrina that shut down oil production in the Gulf of Mexico in the summer of 2005. Should there be a crisis, at the same time, in several of these regions, the supply of the global market with oil and gas could indeed be seriously endangered. The knowledge of the possibility of such a coincidence is enough to keep prices at a high level. Iran's fundamentalist president Ahmadinejad exploited this knowledge when he announced that the oil price could rise to US$200 per barrel at the time of the war in Lebanon. In addition, when Venezuela's populist President Chavez threatens once again to stop exports to the US, he focuses on the psychology of international oil traders. Even gas superpower Russia flirts with the threat of diverting its exports from Western Europe to East Asia in future. Although such a threat is technically difficult to implement, and would only be realistic in the long run, it does have an impact on the recipient. The feeling of being dependent on Russian gas has already changed political relations between Russia and the West in Moscow's favour. Both European and East Asian states are currently trying to sign long-term supply contracts with Russia. To ensure security of supply over an extended period, they are prepared to accept contract conditions above the current price on the global market.

The term 'energy security' has a different meaning for different countries. For Europe and North America, it still means – true to Churchill's principle – the necessity for diversification and access to different energy suppliers and sources. This is at least how the 2006 Energy Green Paper of the European Commission defines the term. China and India fear that their rapidly growing economies will run out of energy and that they will therefore be unable to reach their full development potential. Japan, which has to import all fossil fuels by sea, is first and foremost interested in open global markets and the possibility of being able to invest freely anywhere in the world. The US, also, prefers a liberalized world energy market without any state-imposed restrictions. Already today, American energy companies play a dominant role in worldwide oil trade and want to extend this further. On the other hand, when Russia, Saudi Arabia and the other major oil and gas exporters speak of energy security, they refer to the securing of reliable consumer markets.

To ensure stable global energy markets in the long run, the production as well as the transport and processing capacities must have backup margins. Otherwise, the system as a whole is not crisis-resistant. The more an energy system is structured in a decentralized way, the less pipelines and waterways need to be guarded, and the lower the chance that a terrorist act in the Strait of Hormuz could cause share prices on Wall Street to collapse. If renewable energy sources result in greater independence from imports and a more versatile energy supply structure, they may offer an improvement in energy security, provided that crude oil from Saudi Arabia is not simply replaced by biofuels from Brazil.

Transparency and free access to reliable information on production quantities and reserves are important. Only then can price speculation be prevented and the suspicion be refuted that production and export of certain energy sources are artificially kept at a low level to keep prices high. The Western industrialized countries therefore encourage the important emerging markets, such as China and India, to join the International Energy Agency (IEA) over the medium term. The IEA collects and publishes internationally comparable data on energy production and trade, thus providing a reliable basis for energy policy decisions of states and private companies. In authoritarian states, such as Russia and China, economic data, such as energy consumption and local reserves, are often guarded like state secrets.

A reliable energy supply is one of the basic prerequisites for a functioning economy. Today, in times of economic globalization, only a few countries succeed in staying absolutely independent of imports for their energy supply. These few exceptions include small countries that are particularly favoured by their natural resources. Iceland, for example, can fulfil its energy demand almost completely with local hydro- and geothermal energy. In the long run, Iceland even intends to replace its crude oil imports by liquid hydrogen produced from renewable sources. Unlike Iceland, most other European countries are neither small islands, nor do they have the natural resources to ensure self-sufficient energy supply. When it comes to energy, most countries depend on cooperation with their neighbours and other more distant countries.

During the industrialization era, the development of a national energy industry was directed by the government in most countries. Today, a mixture of market and state, of public and private supply companies has developed in all Western countries. The member states of the European Union (EU) decided in the 1990s to create a joint national energy market and to liberalize their energy markets. Since then, numerous formerly state-owned energy companies have been privatized, for example, in Germany and Great Britain. In addition to this, the national energy markets within the EU are gradually being opened up to companies from other member states. Economic integration can lead to political integration, as is the case in the EU. Mutual dependences create the incentive to solve problems within common political institutions. Economic self-sufficiency is no longer possible in a globalizing economy, even in terms of energy supply. Resource nationalism and attempting to leave the global market do not work. The political control of the energy markets should therefore be exercised through international cooperation and treaties. A country's individual wealth in oil and gas should not become an instrument of authoritarian national policy or imperial foreign policy.

Critical energy infrastructure

Energy infrastructure is one of the most vulnerable elements of modern societies. After the terror attacks of 11 September 2001, people realized how quickly an attack carried out using relatively moderate technical means can damage the economic foundation of the greatest industrial power in the world. Air traffic in the entire US came to a standstill for a week. It took two years to completely repair the transport links in the metropolis of New York. The Wall Street stock exchange, situated close to the World Trade Center, was closed for several days – last but not least because access roads and data lines had been severely damaged. The energy infrastructure of modern industrialized societies is at least equally as vulnerable as roads and data lines. What would happen if the next attack is directed against one of the major international oil and gas pipelines?

Whether by sea, land or air, nuclear power stations and nuclear materials transport are especially vulnerable to terrorist attacks and military disputes. After 11 September, there were concrete indications that terrorists had also planned attacks against Western nuclear plants. As a consequence, the nuclear power stations in the US and some Western European countries were shut down for several days or even weeks.

New Year's Day 2006 was an illustrative example of how our seemingly secure oil and gas supply could be interrupted. For two former Soviet republics, Georgia in the Caucasus and the Ukraine, the year 2006 began with a gas supply crisis. When Russian supplier company Gazprom stopped gas supplies to the Ukraine because the latter was unwilling to pay the sudden price increase for Russian natural gas, three explosions simultaneously destroyed the most important lines supplying gas from the southern Caucasus in Russia. This interrupted Russian gas supply to Armenia and Georgia. At the same time, acts of sabotage affected some important overland lines in Georgia itself, with the result that the capital Tblisi and other areas throughout the country were without heating during the coldest days of winter. Universities and schools remained closed. Economic life came largely to a standstill, as production plants could no longer be operated. While Armenia, a political ally of Russia, was able to overcome the shortage using its own gas reserves, Georgia could only restore supply for its population after several weeks, and only thanks to some hastily signed supply contracts with Azerbaijan and Iran. To this date, it is unclear whether Islamic separatists, criminal organizations or the Russian secret service were behind the series of attacks. Georgia's government accused the Russian secret service of sabotage with the intention of exerting pressure on the country in the dispute with separatist republic of Southern Ossetia. Fitting this picture is the fact that Russian energy industry plans to buy substantial shares in Georgia. In this way, Georgia would become permanently dependent on Russian state-owned companies and thus the Kremlin.

Western energy consumers were again reminded of the fragile security situation in the Southern Caucasus region when a week-long war broke out between Russia and Georgia for control over Georgia's renegade province, Southern Ossetia. Both the Baku-Tbilisi-Ceyhan oil pipeline and a critical rail connection that carries oil from Azerbaijan through Georgia further west were damaged.

The conflict between Russia and Georgia also has a supraregional dimension. To defend itself against Russian regional dominance, Georgia wishes to become a member of NATO as soon as possible. The US supports the Saakashvili government in this matter. In fact, an important oil pipeline was built with American support from the Caspian Sea through Georgia – past Russia – into Turkey. The European NATO partners are more sceptical; they do not want to get involved in the regional conflicts in the Caucasus and are not in the least interested in a confrontational course with Russia, its main gas supplier.

Oil and gas pipelines often run over long distances through sparsely populated and therefore only poorly guarded territory. Many of the most important pipelines, be they in the east of Russia, the Chinese provinces of Tibet and Xinjiang, in Southern Chad or in the north of Iraq, go through regions of crisis or war. The danger of terror attacks by rebel groups or separatist movements cannot be overestimated. Repairing destroyed pipelines and energy supply lines not only takes a long time, but is also very expensive. Once the damage has been done, it often takes considerable time until the status quo is restored – even after fighting has ended. The Iranian oil industry still suffers from the aftermath of the war against Iraq in the 1980s. In Iraq itself, oil production collapsed after Saddam Hussein was ousted from power and is recovering only with difficulty. During the war in Kosovo, NATO tried to keep Serbia's energy infrastructure intact during air raids. Nevertheless, the destruction of several bridges by air attacks paralysed shipping along the Danube for an extended period. As a consequence, fuel transport by river to neighbouring states, such as Hungary, was also interrupted.

Transport of oil and, in future, liquefied natural gas by sea also holds dangers. During the Iran–Iraq war in the 1980s, both sides regularly shot at tanker ships travelling in the Persian Gulf, even when these flew a neutral flag.

The Strait of Hormuz crossed by oil tankers from Iraq, Saudi Arabia, Iran and Kuwait on their way from the Persian Gulf, and the most important waterway in the world, the Suez Canal, are situated in the immediate vicinity of the troubled Middle East region. More than 80 per cent of Japanese and Korean as well as half of all Chinese oil supplies go via the Strait of Malacca, a strait between Indonesia and Malaysia. The US navy guards all these waterways. Since World War II, the US has virtually replaced the British Empire as the policeman of the oceans. Other countries that also depend on oil imports from the Middle East, such as China, regard this dominant US maritime role with unease.

The European Commission therefore expressed its concerns in a non-published study in preparation for its Energy Green Paper: 'Competition with Japan over oil and gas from the Persian Gulf and Russia will become tougher in future. Japan and China will pay more attention to their maritime supply routes and will deploy their naval forces in the region. India as well will play a bigger role in the competition over these resources and will exercise an influence on the region with its naval forces.'

Reorientation of military strategy

Due to this particular threat, the protection of the energy infrastructure is at the centre of the military strategy of all major powers. The most important oil-exporting country in the world, Saudi Arabia, has been systematically equipped with US military technology over the past few years to be able to defend its substantial oil infrastructure against terrorist attacks. Even after their foreseeable withdrawal from Iraq, the US will continue to maintain troop contingents in the country. It is not by chance that the barrack locations planned to date are near Iraq's major oil production plants. NATO AWACS aircraft monitor the new oil pipeline from Baku, Azerbaijan, to the Turkish port of Ceyhan in the Mediterranean. In addition to NATO partner Turkey, Georgia has also been equipped with modern US military equipment, allegedly even with unmanned military aircraft, so-called drones. A particularly explosive topic, in the case of Saudi Arabia, Iraq and Georgia, is the fact that new weapons are thus supplied to crisis regions, which could be used not only as protection for the energy infrastructure but also for other purposes.

The Bundeswehr, the German Federal Armed Forces, is also increasingly supposed to take over tasks to secure German energy and resource supply and free trade routes. Operation 'Enduring Freedom' adopted by the German Parliament following the attacks of 11 September already provides for military securing of the waterways at the Horn of Africa, one of the most important tanker routes in the world. In the Bundeswehr's draft White Paper on Security Policy, which reached the public in 2005, the somewhat inelegant statements on security policy were: 'Security policy must also target geographically remote regions to counteract any tensions and hostilities between ethnic groups, regional crises, states where organized crime and terrorism escalate, rapidly growing communities that do not offer any prospects for the future. The development and extension of good relations to strategic key states in the different regions, contributions to crisis and conflict management, as well as for promoting regional stability are important fields of action of Germany's security policy.

Because of Germany's dependence on imports and natural resources, it is important in this context to focus particularly on the regions where critical raw materials and energy resources are produced.'

The highly complex energy infrastructure of our globalized economy cannot be defended by military means. Instead, our energy supply systems as well as the general political conditions could be structured so that the risk of military assaults and terrorist attacks is reduced to a minimum. Instead of strengthening the security forces of power plants, positioning paramilitary units along important pipelines, the future energy infrastructure should be designed to be less vulnerable from the very beginning. This means turning away from centralized large-scale technologies towards decentralized networks. Strengthening national energy sources, for example, through the use of renewable energy sources, and diversifying the imports can reduce dependence on a few supply routes. However, a policy that reduces political tensions and promotes regional cooperation is the best reinsurance against the threat for our energy infrastructure.

Blood for oil?

When the US and its allies fought their war against Iraq in 1991, the slogan of peace demonstrations all over Western Europe was 'No blood for oil'. Was the first Gulf War actually undertaken to free occupied Kuwait from the Iraqi army, or was it really about American oil interests? The latter certainly played a role. If Iraq's dictator Saddam Hussein had secured control over Kuwaiti oil production, Iraq would have become the largest oil producer in the world. Besides, the heavily armed Iraqi forces would have threatened Saudi oilfields nearby. This dominance would have been unbearable not only for the US, but for all oil-importing countries.

The Kuwait campaign was not the first war in which the oil factor played a decisive role. When Winston Churchill, the UK's First Lord of the Admiralty in World War I, changed the Royal Navy from coal to oil, he gained an advantage that was decisive for the outcome of the war. The UK could only pursue this strategy successfully because it had discovered large oil reserves in Persia shortly beforehand. At the same time, the British succeeded in sabotaging the German oil supply from Romania.

During World War II, the Russian campaign of Hitler's Wehrmacht was also intended to secure the oilfields in the Caucasus for German warfare. One of the reasons why the Allied Forces eventually defeated Nazi Germany was the Anglo-American sea blockade, which could effectively cut off the German economy from raw material imports and especially oil imports.

During the Suez Crisis in 1956, when former colonial powers France and the UK tried to establish a regime in Egypt that was kindly disposed towards them, and to bring the Suez Canal under their control, the US sided with Egyptian dictator Nasser against its Western allies. Since then, the Americans have taken over the role of a protector of the canal and the entire tanker routes from the Persian Gulf into the Mediterranean Sea. Since the Suez crisis and the end of the former colonial powers, the US has become the guarantor of political order in the Middle East. For a long time, American politics therefore consisted of supporting pro-Western dictators, such as the Shah of Persia and Saudi royal dynasty, and protecting them against Soviet influence. Since the rise of the Islamists and the terror attacks of 11 September, it is evident that this political strategy will lead to destabilization of the region, to a delegitimization of the US in the Islamic world and to the rise of terrorist movements over the long term.

Also the second Iraq War, in which the Americans – with a considerably smaller 'Coalition of the Willing' – removed Saddam Hussein from power for good in 2003, even without a UN mandate, was not fought because of the atrocities of the Iraqi dictator or his alleged weapons of mass destruction, but rather to reorganize politically the entire Middle East under America's control. This new order was to serve two aims: to 'dry out' the origins of terrorism and to keep oil flowing. Since then, a civil war has been going on in Iraq, and oil production has fallen to a lower level than during the time of the embargo that the UN imposed on Iraq after the first Gulf War.

The economic benefit of securing the oil wells by military means is in any case doubtful. The US Department of Energy, for instance, had worked out already before the second Gulf War broke out that if the costs of the military engagement of the US in the region were considered, the oil price should actually be at US$100 per barrel. This is the level the oil price actually reached in January 2008, five years after the invasion.

Energy as a weapon

In today's globalized world, it is obviously impossible to secure one's own energy supply by military means alone. However, what about a strategy of utilizing the control over important energy resources or central elements of international energy infrastructure as a weapon?

Since the first oil crisis in 1973, when the Arabian oil states initially interrupted supplies to the Western world, and then increased the prices, oil has been used as a political weapon. After the American invasion of Iraq in 1991, Saddam Hussein set the oilfields in the south of the country on fire. At present, Iran, in the dispute surrounding its nuclear programme, threatens that it will trigger a worldwide oil crisis. The markets at least take the threat seriously. After every speech, as radical-religious Iranian President Ahmadinejad uses harsher words in the conflict with the US, the price per barrel of crude oil increases, because the analysts at the international petroleum exchanges fear a worsening of the political situation in the Middle East and resulting turbulence on the international energy markets.

Already in 1980, US President Carter, in the context of the Soviet invasion of Afghanistan and the Islamic revolution in Iran, said in his annual State of the Union Address: 'An attempt by any outside force to gain control of the Persian Gulf region will be regarded as an assault on the vital interests of the United States of America and such an assault will be repelled by any means necessary, including military force.' The so-called Carter doctrine was born. Since then, the topic 'energy security' reappears in every new version of the national security strategy of the US as a central point.

The US Ministry of Defense realized long ago that US dependence on oil imports does not increase but rather reduces the security of the country. Not only is the energy supply of the greatest economic power of the world sensitive to price fluctuations, diminishing resources and politically motivated embargoes by the oil states; but the price in terms of foreign policy that the US must pay for securing its global oil supply has continuously risen in the past few years. In political Washington, it has, in the meantime, become dogma that economic dependence on Saudi theocracy, which, after all, produced Osama bin Laden, must be reduced. Besides, securing the oil resources is in constant contradiction to other objectives of US foreign policy, such as the advancing of democracy and containing Russian influence in Eastern Europe and Central Asia. Lastly, the US has entered into a dangerous competition with its rival China for the last oil and gas reserves.

James Woolsey, former Director of the CIA under US President Clinton and now an influential lobbyist in Washington, revived an organization from the Cold War, the 'Committee on the Present Danger (CPD)'. At that stage the organization's function was to uncover communists and to supply propaganda in America's armament competition with the Soviet Union. Today, the CPD has decided to place energy security on the agenda. Woolsey believes that national security requires a completely new energy policy and an economic use of oil. American security is threatened by the US energy imports from radical Islamic regimes such as Saudi Arabia.

It is therefore not by chance that the Pentagon partly financed the latest study by energy savings guru Amory Lovins. The title of the book is *Winning the Oil Endgame* and it describes a Great Game that is completely different from the one that takes place on the geopolitical chessboard of Central Asia. The rules of the game are called innovation and energy efficiency. These objectives are to be achieved by the forces of the free market prevailing over the interests of the established energy multinationals. Lovins thus proves that strategic thinking can be something other than pushing oilrigs and soldiers forwards and backwards on a big map. Influential *New York Times* columnist Thomas Friedman also argues that the energy-political dependence has become the central security question of the US.

The disaster in Iraq has led to a rethinking in wide sections of the foreign-policy establishment of the US. Belief in the central role of military strength has been lost. The Clinton administration used to explain the changes in the world in terms of economic globalization. After 11 September, the Bush administration started to look at the world exclusively through the lens of the war against terrorism. Today, the pendulum is swaying back. Slowly but surely, even the US realizes that the major challenges of global change can only be tackled through international cooperation and a variety of political instruments. If we listen to the arguments of Amory Lovins, Tom Friedman or former US Vice President Al Gore, the most important defence against energy as a weapon is a sustainable restructuring of our energy policy.

The country that pursues its foreign policy objectives by the use of energy as a weapon most noticeably is the new assertive Russia that was shaped by former President Putin. The Red Army has long since withdrawn from Eastern Europe and Central Asia. Instead, Russia's influence in its former dominion is guaranteed through investment by large oil companies in neighbouring countries and through Gazprom's pipeline network. Along with political treaties and the stationing of Russian troops, Russia's energy companies attempt to gain additional influence and market shares in Central Asia and in the Southern Caucasus region. In the countries to its west – Belarus and Ukraine – Russia exerts pressure to procure the oil and gas transit lines that cross them in a westerly direction. Russia wishes to become indispensable as the major energy supplier for the West and as such a global power once again.

Today, China is the second largest energy consumer after the US. However, it has only limited national energy resources to fulfil increasing demand. The Chinese government is increasingly worried that an interruption in energy supply could lead to a weakening of economic growth and, consequently, to social unrest and a threat to the regime, which draws its legitimization above all from good economic development. China responds to this challenge with a wide-ranging international energy strategy. The core of this strategy is to secure direct exports to China by state-owned Chinese oil companies directly controlling the oil production in important exporting countries. Already, Chinese interests collide with those of the US on the global petroleum market. China and the US also compete in terms of geopolitical spheres of influence and supply contracts. State-owned Chinese energy companies are not particular when deciding with whom to do business: China also buys from pariah states, like Sudan, and has signed long-term supply contracts with political antagonists of the US, such as Iran and Venezuela. In the competition with the US, Chinese oil companies advance into Africa and Latin America. China's growing influence is accompanied by diplomatic initiatives that China uses to position itself as a regional counterweight to the US. China's hunger for energy is particularly explosive, from a security point of view, in its own neighbourhood. China wishes to explore oil and gas reserves along its coasts – thus entering into territorial conflicts with its neighbours Japan, Vietnam and the Philippines. An alliance with Russia is designed to secure overland delivery, while delivery by sea is to be secured by reclaiming Taiwan. In all these endeavours, China impinges on American zones of influence.

A new 'Great Game'

At the end of the 19th century, Lord Curzon, the British Viceroy in India, compared Central Asia to a chessboard. He believed that the 'Great Game' for power and zones of influences took place on this board. The players in the 19th century were Russia and the UK. After conquering the Khanates in Central Asia, Russia wanted to advance further south, to Afghanistan and Persia. Great Britain wanted to secure its possessions in India and the sea routes to these, and to stop the Russian advance. After oil had been found in Persia, it was important to secure this new natural resource for the British navy. Eventually, Russia and the UK agreed to divide the Persian Empire into zones of influence. Even now, the colonial experience of that time is an important source of Iranian nationalism as well as its wish to become a nuclear power.

Today, the interests of the great powers collide in Central Asia again. India, China and Russia, but also Iran and the US are involved in the remeasuring of the region, which is all about the energy resources available there. Russia and the US are trying to secure their regional influence through military bases. China and India are investing massively in oil and gas production and the building of pipelines. Iran intends to free itself from its political isolation by becoming a cooperation partner for the growing economies of Asia in their hunger for energy. So far, the EU lacks a strategy for the region, even though European companies invest in the entire Asian region and actively participate in the exploitation of energy resources. Indeed, the EU with its experience in regional economic and political integration could be a model to Asian national states as they wrestle over spheres of influence.

To be able to understand the line-up of the Great Game, it is worth looking at the map of the major oil and gas pipelines that cover the Eurasian double continent like a spider's web. The major threads of this web originate in Russia and are controlled by the pipeline monopolists Gazprom (natural gas) and Transneft (petroleum). Today, Europe purchases most of its natural gas imports and a growing percentage of its petroleum via this Russian network. China and the other countries of East Asia also wish to benefit from Russia's reserves in the near future. Both parties, Europeans and Asians, fear that their dependence on the Russian transport monopoly will increase further in future. They are therefore in the process of constructing alternative pipeline routes. Even though the

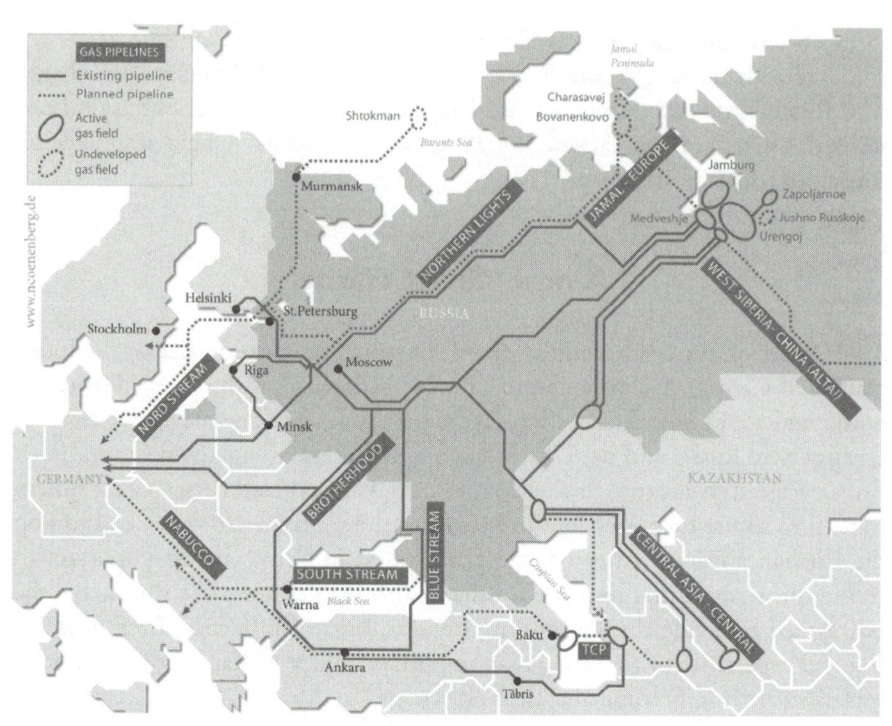

Figure 2.1 The main gas pipelines in western Russia

US presently imports hardly any oil and gas from Russia, they watch Moscow's growing regional influence with unease. Washington also wishes to secure access for American companies to the energy reserves of Central Asia and the Caspian region. Consequently, the Americans support the desire of Central Asian states to free themselves from the political and energy grip of Russia. At the same time, the US is trying to counter growing Chinese influence. And it wishes to isolate Iran. However, a glance at the map shows that the number of available land routes for new pipelines that bypass both Russia, its allies and Iran is restricted. For this reason, the US is not particular when choosing its allies and has done business with dubious regimes, such as the Alijev government in Azerbaijan and the autocratically governed Kazakhstan.

The two central projects of America's policy in Central Asia are the Baku Tblisi Ceyhan (BTC) pipeline and the alliance with Kazakhstan, as mentioned above. The BTC pipeline has been built, transporting oil from the Caspian region by ship to Azerbaijan's capital Baku and from there via Tblisi in Georgia to the Turkish harbour of Ceyhan on the Mediterranean Sea. Another transcontinental gas pipeline is to go from the east of Kazakhstan via Uzbekistan and Turkmenistan, beneath the Caspian Sea through Southern Caucasus and from there to the West. However, Kazakhstan also leaves open its options of an alliance with Russia. On the same day that US Vice President Cheney visited the country in May 2006, the national pipeline operator signed an agreement with Russia and China for exporting oil from western Siberia to China. India also wishes to strengthen oil imports from Kazakhstan and neighbour Turkmenistan. The transport would go through Afghanistan, which is still controlled politically by US and NATO troops.

The energy and security interests of the US in the region intermingle. Since the terror attacks of 11 September 2001, the US has built a network of military bases in the states of Central Asia. Georgia in the Caucasus is even to become a member of NATO. The official motivations for the growing regional engagement of the US military are Islamic terrorism and military deployment in Afghanistan. However, it also lays out the geostrategic territory vis-à-vis great power competitors Russia and China.

Whereas the economy is booming in the states of Eastern and Southern Asia, economic and political crises in the former Soviet republics of Central Asia continue. Turkmenistan and Kazakhstan have developed into major oil and gas exporters. Economic modernization, on the other hand, has halted. Behind the apparently peaceful facade of these authoritarian countries, the danger of political extremism is growing. In Central Asia, all countries try to play off the interests of Russia, China and the US against each other. Kyrgyzstan, for instance, accommodates a Russian and an American military base. The two bases are only 25km apart.

The contradictions in America's policies in Central Asia can be illustrated by a visit of US Vice President Cheney to the region. In May 2006, Cheney travelled to Eastern Europe and Central Asia. In Zagreb, he met the ministers of defence of Croatia, Macedonia and Albania, promising them that they would soon become members of NATO. The extension of NATO, in particular, by including states of the former Soviet influence zone, is the declared objective of the US. On the one hand, the US wishes to extend NATO from a merely transatlantic organization to a global alliance of democracies; on the other, this goes along with a strategy of geographic containment of its competitors Russia and China. The energy factor plays a key role. From Zagreb, Cheney travelled on to Vilnius. In the capital of Lithuania, he gave a speech on democracy in Eastern Europe, criticizing Russia's foreign policy. He accused Russia of using its oil and gas exports as 'tools of intimidation or blackmail, either by supply manipulation or attempts to monopolize transportation'. Cheney was arguably correct in this matter. However, his plea for more democracy and free markets contrasted noticeably with the next stop of his trip, a visit to Astana, the capital of Kazakhstan.

Kazakhstan is the key country for the economic and political future of Central Asia. The president of Kazakhstan, Nursultan Nasarbajev, was re-elected in December 2005 with 91 per cent of the votes cast. Since 1989, he has headed the country uncontested. Before that, in the Soviet era, he had been a top-ranking office holder in the Communist Party. There was little talk of democracy during Cheney's visit, because Kazakhstan has oil. Kazakhstan is a secular Muslim country in a difficult foreign policy environment and with a deteriorating human rights situation. But Kazakhstan is also one of the world's richest countries in natural resources. Most important are the large oil and gas fields under the Caspian Sea: in 2005, Kazakhstan produced 1.2 billion barrels of oil. In 2010, Kazakhstan intends to overtake oil-exporting giants Kuwait and Nigeria. By 2015, production is to be increased to 3 million barrels. In the future, Kazakhstan's oil exports will predominantly go to China. China's role as competitor for the oil reserves in the Middle East will thus be eased.

Kazakhstan's foreign policy is open to all comers. Kazakh oil companies invest in Eastern Europe and in neighbouring Asian countries. There, they also compete with Russia. China and India, but also the West, are interested in its oil and gas wealth. Even though Russia does not need oil from Kazakhstan, it plays a key role for export because it controls the Central Asian pipeline network. The US at all costs wishes to avert an economic alliance, but above all an energy alliance, of the two great energy powers Kazakhstan and Russia.

Where does Europe stand?

Can Europe keep up in the great game for worldwide energy resources? Should Europe play according to the same often brutal rules, or should it opt for a different way of achieving a secure, affordable and environmentally friendly energy supply?

Since the Treaty of Maastricht established a Common Foreign and Security Policy (CFSP), the EU has developed, addressing problem after problem. The target of sustainable development, also in terms of European foreign policy, was written into the Maastricht Treaty. Climate protection and energy security were named as challenges for the EU security strategy of 2003.

Geographically, EU foreign policy grows in concentric circles. Nearby, in Eastern Europe and in the Mediterranean, the EU is already the decisive factor. From a worldwide perspective, however, the EU has only recently emerged from its political childhood. Today it is a global power in international trade policy. After the US, China and some major developing markets, the EU is the most important player within the World Trade Organization. In only a few years, the euro has developed into the second global reserve currency alongside the US dollar. In the not-too-distant future, the moment will probably come when global oil transactions are quoted in dollars and in euros. The EU also plays a leading role in global climate policy, a central component of energy policy. Because of its economic power, its political influence and the responsibility for using it are also on the increase.

Whereas the EU failed in the Balkans in the 1990s, namely by not assuming a leading role in conflict resolution in its own vicinity, it should make good by bringing the Eastern European states towards democracy, security and stability. Russia is needed as a constructive partner in this context. Where the Russian government does not wish to take responsibility, or where it holds misdirected ideas of a Russian sphere of influence for those other rules in force in Europe, the EU must act against the wishes of Moscow and in the interest of its member states and the democratic movements in its neighbouring countries. In this context, the transformation of Belarus, the last dictatorship in Europe, towards democracy is the most important touchstone for EU–Russian relations.

The second cornerstone for these relations is the cooperative structure of the Pan-European energy market. Unfortunately, the relationship between the EU and Russia is currently primarily defined by mutual dependence as far as energy policy is concerned. Russia's foreign policy is increasingly implemented using instruments of economic policy, the most prominent of these being energy exports and investments. Other important aspects of mutual relations, such as the security challenges mentioned above or a widely based economic relationship, take a back seat. An energy policy for all of Europe, as already provided for in the European Energy Charter, must consider the economic and political interests of exporting, importing and transit countries. The obligations in terms of climate protection, to which the EU and Russia agreed when they signed the Kyoto Protocol, also form part of the further development of the European energy market.

The issue of future energy security clearly shows how closely the interests of Europe and East Asia are intertwined in the Eurasian region. The economies of East Asia define their relations to their previously largely neglected neighbour Russia primarily by expected oil and gas imports. The EU and the emerging East Asian economies therefore increasingly compete with each other. This must be structured in a positive way through the increased engagement of the EU in the region.

The further the global range of the common foreign and security policy of the EU develops, the more important it is to coordinate European policy with Europe's most important partner in the world – the US. Even if there were a conflict of interest between the EU and the US, the major global challenges and most regional security conflicts can only be solved if the EU and the US cooperate constructively.

What should be the institutional framework and what should be the contractual basis for such cooperation? The United Nations must remain the most important framework for solving international problems. For this purpose, it must be reformed with the double objective of more efficiency and greater legitimacy for its actions. After the undignified dispute between Germany and other

EU partners over a permanent German seat on the Security Council, forces should focus on producing a common EU position before important decisions are made and – based on its experience of constructive cooperation – a common EU seat should be secured before long.

Chapter 3

Energy Superpower Russia

On New Year's Eve 1999, when champagne corks were popping to celebrate the arrival of the new millennium and television screens were showing pictures of fireworks in all the capitals of the world, one report in the international news column attracted little interest. Russian President Boris Yeltsin had unexpectedly resigned and had handed his official duties over to his Prime Minister Vladimir Putin who, until then, had not been very well known either at home or abroad. On the morning of New Year's Day 2000, the big computer crash had failed to happen, the Earth was still turning, but for Russia, a new era had begun. In addition, as became evident later, Putin's rise led to a turning point in worldwide energy policy.

Russia's comeback

Russia plays a key role in the future energy supply for Europe and Asia. Therefore, it is important to understand where Russia is heading and who the men in control in the Kremlin are, how they think and what they want. Developments in Russia's domestic policy also influence its conduct towards its neighbours and the rest of the world. As was the case during the Soviet era, autocratic domestic policy and imperial foreign policy always go hand in hand. This time, the new great power politics of Russia focuses on the power of Gazprom, not on the weapons of the Red Army. The EU must decide how it intends to deal with this uncanny colossus. Will Russia become a strategic partner, as is always stated in the pleasant-sounding declarations of European summits, or is there a competitor and adversary growing in the East, against whom Europe must prepare itself economically and politically?

Russia's economy is booming. When the Russian rouble collapsed in the late 1990s, solvency of the government could only be saved by international loans. At present, however, Moscow is paying back its foreign debt faster than planned. However, the growth of the economy the past two years is almost exclusively based on the export of natural resources, above all, oil and gas. Otherwise, economic modernization has stagnated. The exploitation of new production fields for these strategically important resources is driven by the closely coordinated interaction between the state and major companies. At the same time, the Russian leadership uses the country's key role in supplying energy to Europe and

East Asia to gain back the influence in global politics that it lost when the Soviet Union collapsed.

Seemingly limitless resources

Russia has the largest coal, uranium and gas reserves as well as the seventh largest oil reserves in the world. In terms of hard coal units (HCU), a widely used unit for measuring energy content, the country has the second highest percentage of fossil fuel overall after the US and far ahead of India, China or Saudi Arabia. Due to the chaotic privatization policy pursued during the time of Yeltsin, when formerly state-owned companies frequently fell into the hands of private financial speculators who neglected production and were only interested in quick resale, and due to the collapse of the consumer markets, production of oil and gas went down dramatically following the end of the Soviet Union at the beginning of the 1990s. Today, it is rising again just as quickly. However, the upward growth potential for Russia's booming energy sector is not without limits.

Figure 3.1 Oil and gas fields in Russia

Russian oil production began at the end of the 19th century in the Caspian Sea in what is today Azerbaijan. Now the major production rigs are found in the west of Siberia. For the future, Russia relies on new oilfields in the east of Siberia, in the Barents Sea and on Sakhalin. Oil experts make a distinction between 'reserves' that are immediately accessible today on the one hand and 'resources' that will only be available in the future. In terms of the reserves of petroleum that can currently be produced in an economical and technically feasible manner, Russia ranks seventh in the world. As far as the resources are concerned, Russia, with 16 per cent, ranks second worldwide. Even though Russia has significantly less crude oil reserves than the number one, Saudi Arabia, its current production volume is almost the same. In 2005, for the first time, Russian oil production exceeded that of the Saudis. However, Russia consumed so much oil at home that its export quantities still lagged behind significantly. Currently, Russia produces almost 5 per cent of its reserves every year. Even if new deposits are discovered, Russia's oil production has already peaked. In future, exports can only increase if Russia itself manages its energy more economically and reduces consumption at home. For this reason, Russia must urgently invest not only in modern production systems and new pipelines, but also in the energy efficiency of its industry and households.

Natural gas production in the west of Siberia, where Europe currently obtains its supplies, is continuously declining. The gigantic new offshore gas field in the Arctic, or on the Peninsula of Yamal, can only be exploited if Western investors contribute to the enormous cost. Oil production in the Caspian Sea and in Central Asia, on the other hand, continues to rise. Russia is therefore interested in directing the export from Central Asia via its pipeline network towards the West, to earn a share in this business. Besides, Russia intends to export liquefied natural gas by tanker to the US and East Asia in the future. Nevertheless, Russia's oil and gas reserves are not limitless, as Russian politicians and Western media sometimes wish us to believe. Furthermore, Russia's oil will run out sooner than her natural gas.

Model Gazprom

Russia's former President Putin, as well as his successor Dimitry Medvedev, does not want to revert to the Soviet era. Putin's idols are not Stalin or Brezhnev. Over his desk in the Kremlin, there hung a picture of Peter the Great, the tsar who undertook the modernization of his country; so far, Medvedev has not replaced it. As Tsar Peter once did, Putin and Medvedev wish to lead Russia to new greatness by adopting Western methods and technologies without transforming it into a Western country. The new Russia has adapted the concept of a major company that is competitive on the global market to the Russian situation. The objective of these new companies is not only to make money, but also to serve the Russian state and its imperial foreign policy. For such purposes, key industries that were privatized during the time of Yeltsin are to be returned to state control.

In the 1990s, Russia belonged to the oligarchs. After they had become rich thanks to wild privatization, private oil companies Yukos, Lukoil and Sibneft entered the global market. State-owned Gazprom, managed by Yeltsin's former Prime Minister Victor Chernomyrdin, looked like a fossil from a different era. The facilities and pipelines were not modernized and noticeably dilapidated; investment was neglected, postponed or simply did not happen. The national price for gas sold to Russian companies and households was highly subsidized. The money from lucrative foreign business disappeared into dubious hands.

Today, the Kremlin wishes to turn Russia's energy companies into competitive players on the global market. However, these companies need more money for their aggressive acquisition strategy. Even if oil and gas prices go up, their own funds are insufficient for global expansion. The only way to get the money is from the international capital market. So far Russia has allowed foreign investors only minority shares in strategically important industries, such as the energy sector. As a matter of principle, the majority of the shares and thus the political control over the course of the company remain with the Russian state. If Russia adheres to this strategy, its major companies will not have sufficient capital for the challenges they face. If the Russian economy opens up completely to the global capital market, the companies will also have to change. Non-transparent state-owned companies that must now serve the political interests of the Kremlin and not their investors would become multinational companies governed by economic considerations and shareholder value. At the same time, however, the Kremlin would relinquish the most important instrument of its current foreign policy. For some time already, politics in Russia has been balancing this tension between benefiting from economic globalization on the one hand and the wish to maintain the Kremlin's influence over strategically significant industrial sectors on the other. As long as the oil price stays at a high level and everybody wants to do business with Russia's companies, the Kremlin succeeds in combining the two objectives cleverly and attracts international investors while maintaining legal and political privileges for its state-owned companies.

Gazprom, which was created from the former Soviet ministry of oil and natural gas, is not only the fifth largest commercial company in the world, but also a state within the state. What has been said about car manufacturer General Motors in the past could be rephrased for the Russian context: what is good for Gazprom is good for Russia. In any case, if we wish to understand Russian politics, we must learn to understand the central role of Gazprom.

Today, Gazprom has become a major global company that is managed using the instruments of capitalism for the benefit of the Russian state. With a 20 per cent share and rising in the global market, Gazprom is the biggest gas company on the planet. The company contributes 8 per cent to Russian economic output, pays one quarter of the entire tax income and supplies half the industry in the country with energy.

The Russian state holds 51 per cent of Gazprom; the remaining shares are held by private owners, some of which are not known. The group has 250,000 employees and maintains a network of 175 subsidiaries as well as a number of holdings at home and abroad that are used to channel billions into economic investment and private pockets. Gas trade with Ukraine and other Eastern European states, for example, is done via a number of intermediaries of whom we know little. RosUkrEnergo, an intermediary trade company that exports natural gas from Russia and Central Asia to Ukraine and other Eastern European countries, is registered in Switzerland. Fifty per cent of the company shares are held by ARosgas Holding AG, a Gazprom company that is also registered in Switzerland. The remaining 50 per cent are held by a company with the name Centragas Holding. It is not known who the owners of Centragas are. However, rumours are circulating repeatedly that known Russian politicians and the Ukrainian mafia are behind it. Recently, the Ukrainian government under Prime Minister Yulia Timoshenko decided to limit RosUkrEnergo's role as an intermediary between Gazprom and Ukraine's state gas company.

Gazprom holds the largest gas reserves in the world and owns 116 billion barrels of oil. It became an oil company by systematically buying out a few smaller oil companies. Its most important unexploited gas reserves are in the Shtokman gas field in the Arctic. The licence to explore for and produce gas and condensate on the Shtokman field is owned by Russian company Sevmorneftegaz, a wholly owned subsidiary of Gazprom. US companies Chevron and ConocoPhillips, Norwegian NorskHydro and StatoilHydro, and French Total applied for the rights to participate in gas production in the north of Russia. In late 2006, it decided to start exploiting the Shtokman field using its own funds and without foreign investors for the time being – an indication of how well filled Gazprom's war chest is. However, the announcement was just a new move in the game to ensure better conditions from Western business partners. Now, the newly incorporated Shtokman Development Corporation has been registered in the Swiss canton Zug, a tax haven that also hosts RosUkrEnergo and numerous other international energy and trading firms. Gazprom owns 51 per cent of its shares, while Total holds 25 per cent and Statoil Hydro 24 per cent. Chevron finally gave up its bid after the Russian government had signalled that US capital was not wished for in this strategically important project.

In the negotiations over valuable Russian oil and gas production licences, the Kremlin always aims to achieve the highest price and political concessions. Access to the Shtokman gas field is therefore part of the negotiations over Russia's joining the WTO.

Other joint ventures of Gazprom with international partners are on the Pacific Island of Sakhalin and in Juschno Russkoje in Siberia. German companies E.ON and BASF-subsidiary Wintershall are also involved in these. However, the majority of the shares in each of these companies remains with Gazprom and thus under the control of the Russian state.

Gazprom ceased being merely an energy company long ago. A state monopolist without any competition at home, the company was in a position to invest its profits massively in other industries. In the period 2000–2005, Gazprom spent US$32billion investing in the tourism, agriculture and media industries. The company owns important Russian daily and weekly newspapers, such as the renowned *Izvestia*, as well as TV stations. As the Russian state is the majority shareholder of Gazprom, the Kremlin also exercises control over the Russian media landscape in this manner. Recently, Gazprom has begun building its own ski resort in Krasnaja Poljana. The goal is to support the Olympic Winter Games in 2014, thus earning further glory for the Russian state.

The downside of this major shopping trip is that Gazprom has fewer funds to invest in the exploitation of new gas fields and the upgrade of its infrastructure. The Carnegie Center in Moscow, one of the most important independent think tanks in the country, for instance, estimates that the exploitation of the gas fields on the Peninsula of Yamal in the Arctic alone will require investment amounting to US$100 billion. If development of these new gas fields were delayed, Russian production would decrease rapidly in the next few years. Here, the company increasingly depends on the support of foreign investors.

For market economy oriented reformers in Russia, Gazprom is the decisive test that will reveal whether this reform to a modern market economy is successful. The price system of the company still stems from Soviet times. Two parallel price systems are in place for international gas exports: the Russians charge the global market price for gas exports to Western Europe; exports to political allies, such as Belarus or Armenia, are invoiced at a reduced price. In some cases, this political price is kept low by adding cheaper natural gas from Turkmenistan. The Turkmen have no choice but to participate in this game: they need the Russian pipeline network for their exports to the global market. For the national market, different gas price subsidies are in place, applicable to both private households and industry. Russia's social system and thus the domestic situation have only remained stable during the past few years because low salaries and pensions were supplemented by a number of state grants. Gas and electricity are almost free for private consumers. As a result, people still open and close the windows to control the temperature in their apartments. Investment in modern systems and control technology are not financially feasible under these conditions. Besides, energy price subsidies for Russian industry were one of the major obstacles to the country's joining the WTO. Even though the other WTO members are willing to accept the reduced prices for private consumers as social grants, they reject any state subsidies to companies they have to compete with on the global market. The EU and the US have, in the meantime, agreed to Russia becoming a member of the WTO. Access to Russia's resources and the Russian market is too important to them. Interestingly enough, only the two former Soviet republics of Moldova and Georgia still need to agree to Russia's application for membership. They, in turn, use their approval as a negotiation factor for low gas prices and further economic concessions.

As Gazprom actually only earns money on the global market and makes losses at home, the income of the company remains below its potential, the victims being not only investment in its own infrastructure and exploitation of new fields, but also the possibility of buying into energy distribution systems abroad. For the future, Gazprom aims to have direct access to its Western customers. For example, the company intended to buy shares in British gas supply company Centrica, which the British government successfully averted. One reason for this is that Gazprom's reputation as a reliable business partner has been in tatters since the head of the company, Miller, closed the gas tap supplying Ukraine in early 2006 with cameras running.

Gazprom is therefore interested in improving its image, at least in Germany, and has, in the meantime, become the sponsor of the highly indebted German Bundesliga (premier league) football club Schalke 04. As early as the beginning of the 1990s, E.ON and Wintershall founded the company Wingas for the distribution of Russian natural gas in Germany. On the other hand, Western European companies wish to invest in the booming Russian energy sector. So far, any involvement is only allowed up to a maximum of 49 per cent. German companies E.ON and Wintershall, for instance, hold 24.5 per cent in the consortium for the construction of the controversial Baltic Sea Pipeline, while Gazprom continues holding the strings with its remaining 51 per cent.

Rosneft, Lukoil and partners

Most Russian energy companies that made the transition from wild privatization at the time of Yeltsin to the state-monopolist capitalism of the Putin era work according to the Gazprom model. The majority shares in Siberian oil company Sibneft, which was once owned by billionaire Roman Abramovic, has, in the meantime, been bought by Gazprom and has thus been virtually renationalized. Abramovic had become rich through Sibneft and invested his money abroad. Among others, he owns the English Chelsea Football Club in London. However, the former oil oligarch remains the governor of the province of Chukotka in the Arctic and, as local representative of Moscow's interests, makes himself popular with the men in the Kremlin. Michail Chodorkovski, the founder and main shareholder of the oil company Yukos, on the other hand, had higher political ambitions, allegedly even to become President Putin's successor. Yukos has, in the meantime, been pushed into bankruptcy by the Russian state authorities. Chodorkovski has been sentenced to several years in prison.

In 2005, 70 per cent of Russian oil production was still held nominally by private owners. Foreign companies held 10 per cent of the shares. Access to the 'sanctum' of the Russian economy, as President Putin puts it, is strictly regulated. Therefore, the Kremlin is always present at the table whenever sales negotiations are held. Besides, the state's indirect control of the oil sector through authorization of production licences for the state-owned oilfields, the tax regulations and the pipeline monopoly goes even further than its direct control by the – compared to the gas sector – relatively low share in the Russian oil industry. In the Yukos case, Russia's Byzantine tax regulations were applied until the once largest oil company of the country had to file for bankruptcy. The selective application of environmental regulations for disciplining insubordinate business partners has also proved successful. The Yukos company had been accused of endangering the catchment area of Lake Baikal with its planned pipeline to China. Shell's production authorization for the largest foreign investment in Russia, the oilfield Sakhalin 2, was withdrawn after an inspector from the Russian environment agency noticed two dead fish floating in the river next to the construction site. The real background was that Gazprom wished to buy shares in the profitable project against the wish of majority shareowner Shell. Russian and international non-governmental organizations (INGOs) had already pointed out the real environmental problems at Lake Baikal and at Sakhalin and were consistently ignored by the environmental authorities of the state.

This is also the reason why the Russian state is not interested in renationalizing the energy sector completely. Instead, a symbiosis has developed in that important officials of the Kremlin simultaneously hold high positions in private business. It is thus guaranteed that Russia's private enterprises also orient their business conduct towards the political interests of the government. The problems of Yukos and its owner Michail Chodorkovski began when the latter tried to free himself from the grip of the state and to sell parts of the company to American oil companies. Yukos' headquarters were to be moved completely to London. Chodorkovski wanted to turn Yukos into a multinational group with an international owner structure. At the same time, Yukos intended to buy its own private oil pipeline to China, thus breaking the export monopoly of state-owned Transneft.

By breaking up Yukos, the Kremlin succeeded in keeping this centrepiece of the Russian oil industry under the control of the state. The television images that showed Michail Chodorkovski, once the richest man in Russia, sitting in a cage in the courtroom, made the other oligarchs in the country shiver. In the near future, none of them will try to flee the spell of the Kremlin or get involved in politics. Chodorkovski's foundation 'Open Russia', where part of the private money of the fallen oligarch was safely invested before Yukos went bankrupt, is still active today. Open Russia supports democratic and human rights organizations as well as its own think tank, the EU-Russia Centre with headquarters in London. As is the case with the Open Society Foundation of former financial speculator George Soros or American foundations, Open Russia can only operate with difficulty within the country itself. However, private foundations that promote the development of a democratic civil society hold the key for a gradual democratization of Russian economy and politics.

Today, however, Rosneft not Yukos is the model for the Russian oil sector. Even by the standards of the industry, state-owned Russian oil company Rosneft is growing at breathtaking speed. Rosneft was also established from the former Soviet ministry of oil and gas in the mid-1990s. Over the past few years, the company has systematically purchased small private oil companies, plants, production licences and a share in Gazprom. Its major achievement was in 2004 when Rosneft, following a number of non-transparent financial moves, secured for itself the oil company Yuganskneftegas, the most important part of the Yukos group. Following a transaction that British business magazine *The Economist* described as 'surreal even by Russian standards', the centrepiece of Yukos was first transferred to a letterbox company called Baikal Financial Group and then, with the help of Chinese loans, to Rosneft itself.

The successful initial public listing took place simultaneously in London and Moscow in summer 2006 and earned Rosneft fresh capital of almost €10 billion. This money is to be used, among other funds, for repaying loans that Yugansk obtained from Western and Chinese moneylenders. Three international oil companies, British BP, Malaysian Petronas and China's state-owned petroleum group CNPC, bought shares in Rosneft. Their decision was probably based on future access to drilling licences from the state and new exploitation areas, and less on current profit expectations. The Kremlin has made it clear in the past few months that promising foreign investment in Russia's energy industry must be done via state-owned companies Gazprom and Rosneft. Rosneft's Achilles heel is the neglected infrastructure. In future, more investment in new pipelines and production equipment will be necessary. This will also require investment from abroad. Buying individual pieces to build an empire – favoured by political patronage – will no longer be sufficient. The empire will also have to be looked after and maintained.

George Soros, who had become a billionaire through currency speculation, warned investors against investing in a company that had become rich through what is actually the renationalization of the Yukos group. Many Western investors lost a lot of money at that time. Between the lines, however, Rosneft advertises the fact that it is owned by the Russian state and enjoys its patronage. Consequently, Rosneft cannot be expropriated. Rosneft can expect a tolerant basic attitude from authorities when purchasing new production licences, for example, in Eastern Siberia, or for authorizations of new pipelines and plants, and above all when dealing with the fiscal authorities.

A privately owned company, Lukoil is still the largest oil company in Russia, even though Rosneft owns larger proven reserves after it purchased Yugansk.

Over the long term, it appears that Rosneft and Lukoil will divide the work. While Rosneft is going to channel the access for foreign investors to the Russian market, Lukoil will increasingly become engaged in international business. For example, Lukoil bought American oil company Getty and thus a network of 2000 petrol stations in the US capital Washington and elsewhere. In the meantime, Lukoil is not only operating in almost all countries in Central Asia, but also in Egypt, Colombia, Iran and Venezuela. Jointly with US company ConocoPhillips, Lukoil is planning the exploitation of the gigantic oilfield in West Qurna, Iraq. ConocoPhillips has, in the meantime, bought a minority share in Lukoil. The Russians also cooperate with China's CNPC. The Kremlin controls politically yet indirectly the privately owned oil company. The warning example of Yukos seems to be rather effective. Lukoil continues to coordinate strategic investment abroad with the Russian government.

In the early 1990s, when the Russian state was almost insolvent and urgently needed foreign funds to exploit its natural resources for export, it gave its permission for several oil and gas production consortia under foreign management, and these consortia were explicitly exempted from a new law that permits foreign companies only to hold minority shares in Russian energy projects. In spite of this, the three major oil production projects on the Pacific Island of Sakhalin (Sakhalin 1 and 2) and in Charjaga in Western Siberia are a thorn in the side of the Russian state, and Russia's energy industry is trying to regain control of these international consortia by reshuffling their shares.

In the case of consortium Sakhalin 1, managed by American Exxon jointly with Japanese and Indian companies, Russian state-owned company Rosneft currently holds only 20 per cent of the shares. The Russians are trying to increase this percentage. Instead of paying cash to foreign partners, the latter are to receive smaller shares in other Rosneft projects in exchange. As a result, Rosneft will hold the majority in every project. To exert mild pressure, the authorizing department is currently denying any extension and expansion of production licences.

In the case of Sakhalin 2, the largest oil production project on the Pacific island to date, the international consortium Sakhalin Energy Investment Company is managed by Dutch-English company Royal Dutch Shell with the involvement of two Japanese companies. For a long time now state-owned natural gas monopolist Gazprom has been trying to obtain shares in this consortium for Sakhalin 2. In return, Shell is to receive Gazprom shares in a gas production field in Western Siberia. In August 2006, the state authority of the environment informed Sakhalin Energy that some environmental restrictions had not been complied with. The consortium's licence was therefore withdrawn.

European and Asian companies that already have operated in Russia for a long time are powerless before this stealthy acquisition strategy. After all, Russia did not ratify the European Energy Charter, which would have protected investors against hostile takeovers. Therefore, ratification of the Energy Charter by the Russian

Duma is top of the agenda at all European–Russian summits and one of the most important European demands on their Russian negotiation partners.

Putin's network

Not only companies, but also people tell the story of the new state-monopolist capitalism that is in the making in Russia. Since he assumed office on New Year's Day 2000, former President Putin has succeeded in establishing a powerful network of personal friends and alliance partners in economic and political key positions. Even with the official transfer of the Russian presidency to his successor Dimitry Medvedev, Putin's shadow still looms large. President Medvedev will probably build on Putin's established network, especially on the powerful clique of businessmen and KGB agents from their home town of St Petersburg. Medvedev though is no KGB alumnus and might therefore bring in a new set of people more closely connected with the energy sector where he last worked. Medvedev worked together with Gazprom chairman Alexei Miller for the City Council of St Petersburg. Before being elected president, he became chairman of the board of Gazprom and Deputy Prime Minister.

In any case, the dominant force in Russia's economy is no longer the oligarchs of the Yeltsin era, who had come to power through privatization profits and criminal dealings, but rather a group of office holders that grew up together with former President Putin in the foreign secret service KGB – today operating under its new name FSB. Even though some of the oligarchs are still in business, they got the message after Yukos founder Chodorkovski was sent to prison that they have to stay out of politics – and any related areas, such as the media.

Whenever high-ranking visitors, politicians or business delegations go to Moscow, a meeting with Gazprom boss Alexei Miller is a fixed date in their diary. In a typical week in April 2006, Miller met the Ambassador of the EU, the CEO of the largest state-owned oil and gas company of India, and American Ambassador William Burns. With the latter, he discussed the involvement of American companies in the exploitation of vast gas fields in the Barents Sea in the Arctic. From there, liquefied natural gas is to be transported by tanker to America.

Gazprom also bought political connections overseas. The consortium for the construction of the new Baltic Sea pipeline is headed by former German Chancellor Gerhard Schröder. His assistant on the board of directors is former Stasi officer Matthias Warnig. Putin, who supported Warnig's nomination to this post, is a former KGB liaison officer in Eastern Germany who knows how to utilize these old connections. Former GDR official for gas pipeline construction Hans-Joachim Gornig manages Gazprom's European sales office in Berlin.

Another colleague of former President Putin from St Petersburg is Igor Sechin. Sechin is the chairman of state-owned oil company Rosneft and at the same time deputy chief of staff in the Presidential Executive Office. Industry Minister Viktor Christenko sidelines as chairman of state-owned oil pipeline monopoly Transneft. Transneft intends to merge with Transnefteprodukt in the near future, Russia's largest company for the transport of refined oil products. Current chairman of Transnefteprodukt is Vladislav Surkow. As Igor Sechin, he also holds the position of a deputy chief of staff in the Presidential Executive Office. Christenko's ministry is currently also working on the establishment of a national aeronautics consortium, which should become a competitor for Boeing and Airbus in the future.

This close connection of economics and politics is typical of companies everywhere in the world. Former politicians are hired by American oil companies as lobbyists to use the old networks to the benefit of their new employer. In today's Russia, the top achievement of a successful political career is to be awarded an executive position in one of the state-owned companies. In the case of Gazprom, the special problem consists of the fact that political contacts of former Stasi and KGB officers had always been secret and difficult to see through for outsiders. It is therefore difficult for the public and the media to understand who has an influence on whom.

However, the choice of Gerhard Schröder is easy to understand from a Russian perspective. Even if it cannot be proved that Schröder had already known about the job offer from the Russians at the time when in office he agreed to the construction of the Baltic Sea pipeline, the former chancellor was the one who made decisive moves in German Russia and in energy policy for years. The story therefore leaves a bad taste in the mouth. Not least because of the Schröder issue, the implementation of a code of conduct is being discussed in Germany that will define rules to govern professional career of former parliamentarians and government ministers. According to these, it will be prohibited for former politicians to join those companies whose direct economic interests they decided while in office.

Politics with pipelines

Apart from Gazprom's monopoly on the Russian gas pipeline network, the Russian state also holds the monopoly on the country's oil pipelines. Today, this monopoly is even statutory. The 100 per cent state-owned Transneft controls 45,000 kilometres of pipelines from Eastern Europe to Asia.

Until the beginning of the new millennium, the majority of Russian oil exports to the West were handled via the Baltic region. Until then, most Russian oil exports went via the port of Ventspils in Latvia. Since a gigantic oil export harbour was built in the city of Primorsk to the north of St Petersburg, Russian pipeline monopolist Transneft is trying to push the Baltic States out of the oil export market. Deliveries to Ventspils were stopped in 2002. Since then, the Baltic States have specialized in the processing of crude oil products, which they obtain from other sources. Deliveries to Lithuania's largest refinery in Mazeikiu were also interrupted. Allegedly extensive repair work is necessary following an accident in the Druschba ('friendship') pipeline.

Gazprom's double monopoly on the national pipeline system and gas export is important, even indispensable for the Russian state for a number of reasons. Through state control, under which even transactions with foreign states and companies fall, the Kremlin maintains its influence on gas export prices, the most important revenue for Russia's national budget. Besides, the construction of pipelines is used systematically for the foreign policy agenda – in this context to secure the sale of and dependence on Russian gas supplies, even later when alternative supplies from Central Asia, North Africa or (as liquefied natural gas) even more remote production areas will be available.

For many years, Russian companies have entered the energy sector of neighbouring countries. Gazprom & Co. thus wish to provide for a time when their own natural resources run dry. The number one target region is the former republics of the Soviet Union and the states of the former Eastern bloc. Russian businessmen and companies still maintain good connections with these. Since Vladimir Putin became the top man in the Kremlin, numerous former members of the foreign secret service have also gained a foothold in the world of business. They bring along their old political contacts on the periphery of the former Soviet empire. What is even more important than personal connections is the common infrastructure shared by former Soviet bloc members. In the Soviet Union, energy supply was planned centrally. Some of the now independent states therefore still depend on energy imports from Russia. The three Baltic states, for instance, purchase 100 per cent of their natural gas from Russia. In addition, the pipeline and railway network, on which coal, oil and gas are transported, connects the former satellite states with Russia. An alternative traffic and energy infrastructure that could link the Baltic states and other Eastern European countries to the EU and thus disconnect them from Russia's influence zone is being created only little by little. The creation of three trans-European networks remains a strategic objective of European foreign policy.

Russia uses the construction of pipelines to make politics with its western and eastern neighbours. Up to now, most Russian natural gas has been transported westwards via Ukraine and Belarus. Other pipelines go via Hungary to the Adriatic Sea and along the edge of the Black Sea into Turkey. Over the long term, the Gazprom group wishes to take over the transit pipelines that go through Belarus and Ukraine in a westerly direction and their respective underground gas reservoirs. Even today, the transit pipelines through Ukraine are managed by a Russian-Ukrainian-German gas transport consortium. Gazprom would like to take over Belarusian Beltransgas, but has so far met resistance from dictator Lukashenko in Minsk. The fees from the gas transit are one of the most important revenues for the state budget with its high deficit. Gazprom will therefore continue turning the price screw until Belarus gives in.

With the planned Baltic Sea Pipeline from Wyborg in Russia to Greifswald in Germany, Russia wishes to secure direct supply for Western markets and reduce its dependence on transit countries Ukraine and Belarus. Later on, the Baltic Sea Pipeline is to be extended into the Netherlands and the UK. The Baltic states and Poland, which were not involved in above-mentioned Russian-Ukrainian-German gas transport consortium, will again be left out of this project. Small countries in Central Europe are not only annoyed because they have no share in the lucrative gas transit business, but they also fear for their own supply security and that they might be isolated and marginalized politically between Germany and Russia.

Russian oil is to be transported to the European market by tanker to the port of Bourgas in Bulgaria and from there via a new pipeline to the Greek port of Alexandropoulos on the Mediterranean. One purpose of this project is to compete with the Baku-Tbilisi-Ceyhan pipeline that Western oil companies use for their imports from the Caspian region, bypassing Russia.

Russia's latest project to compete with the EU's trans-European energy networks (TEN) is the so-called South Stream pipeline that will, if completed, link Bulgaria via Serbia and Hungary with one of Europe's major gas hubs in Austria. Russia has already signed bilateral agreements with Serbia, a traditional ally, and the EU member state Hungary. South Stream will directly compete with the EU-sponsored Nabucco pipeline project. Two trans-Balkan gas pipelines will not be economically feasible. Therefore the geopolitical tug of war between Russia and the EU has started.

Russia is also extending its economic-political influence in the Southern Caucasus, while at the same time trying to prevent any rapprochement of this import transit region with the EU and the US, but also with regional powers Iran and Turkey. Early in 2006, for example, Gazprom took over the Armenian section of a new gas pipeline into Iran as well as the largest thermal power plant of Armenia. Gazprom's share in the common energy supplier ArmRosGazprom thus increased from 45 to 75 per cent. In return, the Russians will continue to invoice cheaper gas prices for the Armenians. In neighbouring Georgia as well, the Russians own part of the energy supply, but not the transit network.

Russia has signed a long-term supply contract over 25 years with Turkmenistan. As part of it, Gazprom buys natural gas from Turkmen state-owned Turkmeneftegas at a fixed price of US$44 per 1000 cubic metres. The price on the global market where Russia can resell the gas is five times the amount in some cases. Turkmenistan is completely dependent on the Russian pipeline network. Besides, protection from Russia is the guarantee for survival for the dictatorial regime of Turkmenistan. This guarantee has its price. The cheap gas from Turkmenistan helps to maintain the subsidized gas prices for Russia's heavy industry and private households at a low level. Producers from Central Asia cannot enter the Western market as competitors. By adding Turkmen gas, it is possible to increase the gas price for neighbouring countries, such as Ukraine, only gradually to world market level. Any sudden increase in the subsidized prices for former Soviet republics would damage Russia's reputation as a reliable energy supplier. Russian companies and Russian capital are also active in Uzbekistan and Kazakhstan. The export networks to the West belong to Russian monopolists in any case.

In Eastern Siberia, several pipelines are planned for the export to China, Japan and Korea. As early as summer 2001, Transneft presented a proposed petroleum pipeline project for the Asian-Pacific region. Since then, discussions have been held on the route and the destination of the pipeline. During these discussions, Russia has cleverly and successfully played off the energy-hungry consumer markets in Eastern Asia against each other. According to the current planning status, the pipeline is to go to the Russian harbour of Nakhodka in the Pacific. From there, the oil is to be transported on by ship to the consumer states.

Russian energy giants have even begun poaching on foreign territories. In June 2006, for instance, Gazprom announced that it would invest up to two billion dollars in the natural gas industry of Bolivia in the Andes Mountains, which had only recently been nationalized by force. Negotiations over compensation with the former investors from Spain and neighbouring Brazil had not even been finalized. Gazprom wishes to become active in Algeria and Libya as well, traditionally the region of influence of French, Italian and German companies.

Even wealthy Western Europe and North America are becoming increasingly important for Russian energy exports. Over the medium term, the US especially is interested in benefiting from Russian energy wealth and reducing their energy dependence on the states of the Middle East. There is already talk that Russia instead of Saudi Arabia is to provide the reserve capacity on the global market and would consequently be in a position to control price fluctuations. As soon as the construction of Russia's first liquefied gas plants has been completed in the oil harbours of Western Siberia and on Sakhalin, natural gas exports to the US can start. Until now, lucrative exports to the US have been prevented by the restricted range of the gas pipeline network. However, the start of liquefied gas production would generate a global market for gas for the first time, with Russia as the dominant player.

Gas OPEC

The building of major companies that are competitive on the global market is not the Kremlin's only interest. Russian energy groups aim not only to increase their own wealth and that of their shareholders, but also to serve Russian industrial and foreign policies. Over the long term, the Kremlin is working to create a 'Gas OPEC'. The Organization of Petroleum Exporting Countries (OPEC) was founded in Baghdad in 1960 as the representation of interests of the petroleum exporting countries. OPEC's member states produce 40 per cent of crude oil worldwide. Their export quota is even higher because numerous other major oil producers, such as the US, consume the petroleum that they produce almost exclusively in their own country. The objective of the organization is to keep production quotas and prices at a constant level. In the 1970s and 1980s, OPEC succeeded in controlling the worldwide market for crude oil. Some of the OPEC members, for instance the Arab oil states, utilized their market power systematically to get political support as well as modern weapon systems from the West. After oil was discovered in the North Sea and Russia entered the global oil market, OPEC's power to influence the worldwide oil price development decreased. Neither Russia nor the North Sea states have joined the organization.

To date, there has not been a worldwide organization of gas exporting countries. At present, discussions are underway in the Kremlin to forge such an alliance from Russia, with its partners in Central Asia and other major exporters. Together with the gas-producing countries in Central Asia and Iran, this group would own a critical quantity of worldwide gas reserves and could dictate prices – as OPEC did in the 1970s. Some strategists in the Kremlin are even speculating about an alliance between petroleum OPEC and a new alliance of gas exporters. The market power of such a worldwide alliance of energy producers would indeed be substantial. Until this bold vision has been realized, many conflicting interests must be overcome so that oil and gas customers do not have to worry.

Russia and Iran alone own 42 per cent of known worldwide gas reserves. The Russian monopoly on the pipeline network on Eurasian territory would guarantee the leading role of the Kremlin within gas OPEC. Even though Iran, Kazakhstan and Turkmenistan might warm to the idea of a price-determining bloc of major gas exporters, they do not like the prospect of falling under the hegemonic influence of Moscow economically and eventually also politically. Therefore, they wish for alternative export routes that bypass Russia. The most important bridging countries in this context are Turkey and the states of the Southern Caucasus. The US is behind the Baku-Tbilisi-Ceyhan pipeline that connects the oilfields in Central Asia with the Turkish port of Ceyhan on the Mediterranean Sea. Gas exports should also go via this route in future. An Austrian consortium by the name of 'Nabucco' – named after Verdi's opera about the Babylonian ruler Nebuchadnezzar – is planning to build a 3400km gas pipeline via Turkey and Bulgaria to Western Europe. China is in the process of connecting its west by a gas pipeline – going parallel to an already existing oil pipeline – to the east of Kazakhstan. The gas export monopoly that Russia now exerts on Central Asia is consequently not going to last. Over the long term, the Russians therefore cannot rely solely on their monopoly but must also maintain close political coordination and price agreements with other regional producers.

What is particularly explosive about the gas OPEC project from a political perspective is the fact that it would create an alliance between Russia and Iran. In its policy regarding Iran, Russia must decide whether it will try, jointly with the West, to prevent Iran's military nuclear programme, or whether it wishes to push back Western and above all American influence in Central Asia together with Iran. Until now, Moscow pursues a see-saw policy as far as Iran is concerned and tries to get as many political concessions as possible from both sides.

On 15 June 2006, Putin announced at a conference in Shanghai that he would support the construction of a new gas pipeline from Iran via Pakistan to China. Iran could then supply less gas to Europe and would no longer be a competitor to Russian gas exports. Using the common pipeline network, Gazprom would also exploit the high-growth Indian subcontinent for itself. From a Russian perspective, Iran is a cooperation partner and competitor at the same time. Just like Russia, Iran is forging ahead on the international oil and gas market. However, because of economic isolation over many years and the US trade embargo imposed on it, Iran is also in urgent need of capital and modern technology. The proposed pipeline project is beneficial for Russia in two ways. Gazprom and the Kremlin believe that the merger of the Iranian and the Russian pipeline network would ensure control over the gas market of Central and Southern Asia for the two countries together. The US will hardly like that. The latter's objective is the economic and political isolation of Iran.

Iran is a key country in the great chess game over political dominance on the Asian continent. Both Russia and China wish to build privileged economic and political relations with the regime in Tehran. Both are trying to bring Iran closer to the Shanghai Cooperation Organisation (SCO). The military nuclear programme is a major obstacle to such cooperation. However, both the Chinese and the Russians are interested in finding a solution that would bind Iran closer to their zone of influence. Both are happy to accept further isolation of this regional power from the US.

The new gas alliance would be similar to the real OPEC in yet another aspect. A new alliance of authoritarian regimes with significant economic power and a menacingly large influence on the Eurasian double continent would be created. Russia, China and the dictatorships in Central Asia support each other whenever criticism from the West and opposition groups in their own countries must be averted. The EU must do everything it can to prevent the formation of such a bloc opposed to its values and interests. One strategic objective of the Kremlin is to stop the advance of Western democracies in Central Asia. To achieve this, Moscow is even willing to support the Islamic regime in Iran that it actually dislikes. However, Russian tolerance would reach its limits if Iran wished to implement its own nuclear weapon programme. In this matter, former superpower Russia is anxious to maintain its privileges.

Atomprom

Following the Gazprom model, Russian nuclear industry is to be united as one effective power under the management of the national nuclear agency Rosatom. In addition to Rosatom, it includes the national power plant company Rosenergoatom, export companies Tenex and Atomstromexport, as well as the uranium trading company TVEL. The new major nuclear company is to be called 'Atomprom' and embodies this new agenda. Atomprom is to benefit from the political renaissance of nuclear energy and should in future be competition on the global market for other international power plant constructors, such as German Siemens AG and Japanese Toshiba group. Russian nuclear power station constructors mainly focus on the Asian market. Apart from the construction of complete nuclear power stations, the Russians can also offer other parts of the nuclear fuel cycle, such as the production of fuel rods and nuclear waste disposal sites. On the one hand, the Russian nuclear industry thus wishes to stand out from competitors on the market that come from other countries with strict environmental and anti-proliferation standards. On the other hand, customers should be tied to the Russian nuclear complex over the long term. This also applies to Iran. Thus, Russia has offered to take over important parts of the nuclear fuel cycle for the Iranian nuclear programme. This is not only to avert the danger of Iran's producing nuclear weapons, but also to ensure a new dependence of their southern neighbour on Russian technology and Russian experts. Some years ago, the Russian government made the same proposition to the Ukraine, which the latter declined with thanks.

Russia also wishes to cooperate with the Chinese in the nuclear area. China intends to construct at least 30 additional nuclear power stations over the next 15 years. For this, they aim to use Russian technology, which Western countries consider outdated and unsafe. The future Atomprom will challenge international competitors, such as American company Westinghouse, on the nuclear market. The most important arguments in favour of Russian suppliers are their low prices and less strict export controls for nuclear technology.

As for the Russian nuclear programme, Europe must choose between two evils. When Ukraine wished to replace the destroyed reactor at Chernobyl in the mid-1990s, it had two alternatives. The EU made Ukraine the proposal to replace the lost capacity of Chernobyl with modern gas power plants. However, Ukraine insisted on replacing the reactor blocks of Chernobyl with new nuclear power stations. Eventually, Kiev threatened that it would continue running Chernobyl and then replace it with reactors of Russian design. After that, the EU declared that it was willing to help with the construction of new nuclear power stations according to Western standards. As far as China and India are concerned, the EU is presently facing the same dilemma. Should it try to get a piece of the cake in East Asia, or should it leave the field to Russian industry?

The West also depends on Russia's cooperation in the fight against nuclear proliferation. In Bushehr in Iran, south-west of Isfahan, Russia is building a 1000 megawatt nuclear power station at an estimated cost of US$850 million. Bushehr will not be the only one. In accordance with an agreement signed in Moscow in 2002, Russia will assist Iran in the construction of at least six nuclear power stations. Iranian engineers and scientists are trained at the Kurchatov Institute for Nuclear Energy in Moscow, and work as trainees in Russian nuclear power stations where they are trained for the operation of the future nuclear park in Iran. After Western cooperation partners, such as Siemens AG from Germany, had withdrawn from Iran, the Iranian nuclear programme was no longer viable without Russian support. Several thousand Russian scientists and engineers are involved in it. The International Atomic Energy Agency (IAEA) and also the EU and the Americans suspect that Iran plans to build an atomic bomb via its civil nuclear programme. Therefore, the UN Security Council was approached – on which Russia has a veto. In addition, the proposal to enrich uranium for Iranian nuclear reactors outside Russia would strengthen Moscow's key role.

Of course, the cooperation of the Russians has its price. At the G8 Summit in Moscow in July 2006, US President Bush offered his colleague Putin a comprehensive nuclear cooperation agreement. Its three main components are: Russian cooperation in the efforts against Iran's nuclear weapon programme, an extensive transfer of US nuclear technology to Russian companies, and the long-term plan to erect an international nuclear waste disposal site somewhere in the vast countryside of Russia.

Countries such as Germany, who decided to phase out nuclear power in their own country, are in a dilemma. If they renounce the export of nuclear technology, other states with often dubious safety standards and a more relaxed attitude as far as the risks of proliferation are concerned, will take over the market. A three-component strategy will be consistent and successful over the long term: the phase-out of nuclear power at home must be continued to prove internationally that a modern industrialized country can function without nuclear power stations; modern and sustainable energy technologies, be they CO_2-neutral coal power plants or renewable energies, must become attractive alternatives to fossil and nuclear energy use on the global market; and those countries that wish to continue using nuclear energy, such as Russia, must be persuaded to adhere strictly to the safety standards of the IAEA. This also applies to the new giant on the nuclear global market, state-owned Russian Atomprom.

On the way to becoming the great energy power

Russia wants to use its seemingly limitless energy resources to become a great power once again. After the collapse of the Soviet Union and many years of economic decline and political humiliation, the strategists in the Kremlin hope to reconstruct the old empire at least economically through its dominion over the largest oil and gas reserves and to regain the world's respect. Former Soviet republics, such as Ukraine, should be made economically dependent on the central power and thus brought to heel politically as well. As long as energy prices rise, there is the feeling that Russia is in a strong negotiation position vis-à-vis the Europeans and Chinese.

As early as at the start of the 1990s, when the Baltic states were striving to gain independence, Russia tried to put pressure on them through an energy blockade. In late 2005, Gazprom suddenly increased the gas prices for the Ukraine, thus creating an emergency for the newly elected Yushchenko government, which the Kremlin disliked. In autumn 2006, Russia's dispute with the Caucasus Republic of Georgia escalated. Here again, Russia threatened to use energy as a weapon. At the beginning of the same year, oil and gas exports had already been interrupted for several months following a pipeline explosion, the circumstances of which have never been clarified. The Baku-Tbilisi-Ceyhan oil pipeline as well as a critical rail connection were both damaged during the Russian-Georgian war that broke out over Southern Ossetia in Summer 2008. Russia has continually tried to destabilize its southern neighbours in the Caucasus to exert leverage over this critical energy transit region. Today, Russia has hardly any neighbour that it has not threatened with energy deprivation as a weapon in the event of any political insubordination.

The nature of being determines the nature of awareness, and with its new economic strength, Russia's view on the world has also changed. The policy of rapprochement with the West, which took precedence under Gorbachev and his successor Yeltsin, is now over. Russia's major companies are to be placed in a position to become competitive on the global market. Nevertheless, opening up to the global economy is not to be accompanied by liberalization at home – in other words, westernization. On this point, the Kremlin has learned from the teachers in Beijing.

The two schools of thought among Russian strategists, the 'Europeans' on the one hand and the 'Eurasians' on the other, are already fighting about which direction the country should take over the long term. The 'Europeans' focus on the country's political and economic integration into the West. Even though Russia should not become a normal member of the EU and NATO, it should have close ties with both organizations and should thus have a say in all decisions of a fundamental nature. Joining the European Energy Charter would be possible and desirable for this group of foreign policy strategists. The best conditions possible for Russia would have to be negotiated.

The 'Eurasians', on the other hand, see Russia's future on the Asian market. Their arguments are: Europe is doomed demographically, economically and politically; the world's future is in Asia; Russia could act as a bridge between the old and the young part of the double continent. Jointly with the other former Soviet republics, it should establish some kind of 'EU East' to comprise the authoritarian states of Central Asia and controlled democracies, such as Russia itself. The alliance with China could be useful for freeing Russia from its unilateral economic dependence on the West.

Unlike China, however, Russia has so far not engaged in a major social modernization strategy. Russia's current economic growth rates, its state budget and thus also its foreign policy weight are based mainly on the export of oil and gas and are thus not diversified. Today, the contribution of energy exports to the gross national product is 21 per cent. The annual inflow of foreign exchange of US$150 billion (2005) fuels the national economy, leads to inflation and a rising exchange rate for the Russian rouble. Investments in other industries and services are therefore no longer profitable; no major modernization of the economy is being undertaken. Driving through Russia today, the harsh difference between the few rich and the many poor, between the wealth island of Moscow and the lack of economic perspectives in the provinces is noticeable. The 'Europeans' will therefore be right eventually: there is no alternative to Russia's rapprochement to the West.

Europe's policy on Russia

For the EU as well, Russia is the most important neighbour. Russia's contribution is indispensable on the way to achieving the goal of creating a zone of security, stability and democracy in Europe as a whole. The major danger that threatens the EU's foreign policy is the formation of an alliance of authoritarian and semi-democratic states in the East. It must confront this danger with all its power. The EU needs a democratic Ukraine and a democratic Belarus beyond its borders. Therefore, relations with the smaller states of Eastern Europe may not come second to the EU–Russian relations and the wishes of Moscow. Only then is there a chance that the European social model can be exported to Russia over the long term. There may not be a separate set of rules for Russia. The democrats in the country expect EU politicians to measure their country by European standards, meaning compliance with democracy and human rights.

Democracy for the neighbours in the East is not only a goal for idealists, a pious wish for the political friendship book. The EU can only tackle the gigantic tasks that lie before it with democratic partners, with partner countries where the principles of a constitutional state govern. One of the major tasks that it faces is a sustainable energy future on the European continent. Here again, Russia is indispensable.

For the near future, the countries of the EU will depend on reliable energy imports from Russia. Even a consistent policy of energy-saving and the substitution of fossil by renewable energies can only completely replace oil and gas over the long term. In the same way, however, Russia also depends on Europe. Without the money from the West, and thus above all from Europe, Russia cannot maintain its energy production, increase its national energy efficiency, or modernize its economy as a whole and prepare for a time when it will no longer be able to live on its natural resource exports alone.

European investments in a sustainable and environmentally friendly energy policy for Russia promise to be beneficial in three ways: for investors from the EU with their substantial know-how, the modernization of Russia's economy, the building of a sustainable energy supply in the cities of the country, and the utilization of the immense potential for renewable energy technologies would be the opportunity of the century, which could also create wealth and jobs in all of Western Europe; and if Russia succeeds in managing its wealth of resources more efficiently, more will be left for export. In all scenarios for a future based on renewable energy, the increasing percentage of natural gas, at least as interim energy, plays a key role in replacing coal and nuclear power. The combustion of natural gas is cleaner and releases less carbon dioxide per energy unit than its fossil competitors coal and oil. For the export of natural gas, Russia could extend its national reserves and could show more concern for the fragile ecology of the new exploitation regions in the Arctic and offshore of Sakhalin. Lastly, the global climate problem cannot be solved without the contribution of Russia.

Now Russia has better cards to play than the Europeans do. However, this is not only because of the high energy prices and because of Europe's increasing dependence on Russian oil and gas, but also because of the lack of an agreement among the Europeans. Europe therefore needs a common energy policy rather than each member going its own way. Clear ideas in energy policy must be one of the pillars of the future Russian strategy of the EU. This includes a comprehensive understanding of energy security.

Chapter 4

The Rise of Asia

China and India joining the global economy has led to a dramatic increase in demand on worldwide energy markets. For these two 'emerging powers' of the 21st century, energy and foreign policy are closely connected. The region and its dynamic economic growth highly depend on energy imports. Apart from oil from the Middle East, energy imports from Russia, Central Asia and Africa play an increasing role. Over the past few years, a close network of energy-political cooperation and trade flows have developed among the states of Central and East Asia.

However, despite this economic cooperation, the emerging powers of Asia compete against each other over access to energy resources. Political tensions have risen in particular between China and its neighbours over the past few years. As most states in East and South Asia have only limited own fossil fuel resources, their focus is on extending nuclear energy. However, in the shadow of civil utilization of nuclear energy, there is a dangerous nuclear armament race. The promotion of renewable energies as an environmentally friendly alternative is slowly gaining momentum. Japan, one of the foremost technological nations in the use of photovoltaic and other modern, environmentally friendly technologies, could lead the way in this regard. Furthermore, any tentative beginnings of regional cooperation have so far been overshadowed by historic rivalries. But closer regional cooperation following the example of the EU would actually be the best alternative to growing political and economic tensions for East and South Asia.

China and India are still trying to establish their role at a global level. Over the medium term, both countries wish to act on a par with the US, Europe and Russia in a multipolar world order. China's state-owned petroleum companies are entering the global market without regard for the environment or human rights. It would be in the interest of Europe if the emerging powers of Asia were to be integrated into an international system of institutions and rules to ensure that their development takes place peacefully.

Asia's new energy hunger

China's economy is booming. According to the government in Beijing, its energy requirement will double in the period 2005–2015 alone. Estimates of the

International Energy Agency (IEA) are more cautious, but also forecast dramatic growth rates.

Until 1993, China was still in a position to export oil. Since then, however, imports have significantly exceeded domestic production every year. In 2003, China was the second largest oil importer in the world for the first time, with more than 6 per cent of the global market share, thus overtaking Japan. The IEA reckons that the percentage of crude oil that China must import will increase from 30 per cent in 2000 to more than 80 per cent in 2030.

Even China's efforts to exploit new oil wells in Tibet, the Western province of Xinjiang and in its coastal waters will not reverse this trend. Besides, China's economy is one of the 'most oil-intensive' in the world. This means that China requires an above average quantity of oil for the production of one unit of economic output. Until now, China has purchased most of its oil from the Near East. In future, oil is also to be imported overland by pipeline from Myanmar, Iran, Central Asia and Russia. To this end Chinese politicians are forging new alliances, such as the Shanghai Cooperation Organisation (SCO), while China's state-owned companies invest in a new, Asia-wide energy infrastructure.

China's hard coal production is by far the largest in the world. According to the IEA, production in 2004 amounted to almost two billion tons and thus 42 per cent of total global production. Chinese hard coal mining, some of which is open-cast, has a considerable impact on the environment. Coal firing in factories and heating systems results in catastrophic air quality in the industrial areas and major cities in the north of China, including the capital Beijing. The Chinese environmental authority estimates that to date the cost of the damage of self-inflicted pollution exceeds the income earned from the annual increase in economic output. The modernization of Chinese coal power plants, including an increase in efficiency, the installation of modern filter systems and the separation of the greenhouse gas CO_2 are therefore among the major topics to be addressed in international cooperation development with China. Over the medium term, the Chinese wish to replace their coal by natural gas, which is cleaner in combustion. Until the end of the 1990s, natural gas played hardly any role in the Chinese energy mixture. In future, however, natural gas is to be imported by pipeline from Russia and Kazakhstan. Domestic production inland and offshore will also play a role. At present, China's nuclear power only provides 1.5 per cent of the total energy requirement. By 2020, it is planned to construct 30 additional reactors and double this percentage.

Insufficient energy supply is increasingly becoming a problem for China's economy. Companies complain about power cuts. During summer, factories in Beijing or Shanghai work only at night because the available electricity is required for running millions of air-conditioning systems during the day. The potential for more efficient energy utilization by using modern systems and renovating buildings is enormous. With the new legislation for the promotion of renewable energy, China wants to start mass-producing solar systems, wind generators and fuel cells. If the Chinese were to be successful with this new technology, they would become a serious competitor to the current world market leader Germany as well as Japan or Denmark in this sector.

In 2004, China contributed 16.5 per cent of worldwide CO_2 emissions, and thus holds the dubious honour of being second after the US. The IEA estimates that China will overtake the US as the biggest CO_2 emitter by 2010. Some estimates saw China already overtaking the US by 2007. By 2050, China's contribution to global CO_2 emissions, which by then will have increased substantially, will be between 25 and 40 per cent. It is of fundamental importance for the future of international climate protection that Asian emerging markets also make their contribution to a sustainable and climate-friendly energy future. Therefore, China, India and other Asian growth economies should be included in the next round of UN climate negotiations, during which treaties will be agreed to replace the Kyoto climate protection protocol, which expires in 2012.

Parallel to China's rise, India, the other country with a population of more than one billion, is also emerging rapidly. According to its own estimates, India's energy consumption will increase 50 per cent by 2015. Like China, the only significant national energy resource that India has is hard coal. At present, 70 per cent of the crude oil demand is covered by imports. In India too, energy scarcity turns out to be a decelerating factor for the booming economy. Energy poverty has become a growing problem for the population of the largest democracy in the world, and a vehemently discussed political topic. Numerous Indian villages are not connected to the electricity grid, and in others power is switched off at night. In India's democracy, access to energy is not only a question of development but also of justice, with 'poor' against 'rich'. Also, in the international context, India, which has one of the lowest energy consumption rates per capita in the world, demands an equal share in the global energy resources. This is also the reason why India resists international commitments on climate protection, even though a more efficient utilization of energy and the use of renewable energy could help in solving the dramatic problem of energy poverty in rural India. Nevertheless, India is one of the few countries worldwide with its own ministry of renewable energy.

In the past few years, India has signed a number of long-term oil and gas supply contracts with other countries in South and Central Asia, among these Iran. Unlike China's state-owned companies with their aggressive demeanour, India mainly focuses on investment in its private sector at home and abroad. India's energy foreign policy is therefore less visible than that of the Chinese. Nevertheless, Indian capital is currently flowing into the exploitation of resources worldwide. Like China, India focuses on nuclear energy for the future and has therefore signed two far-reaching agreements on technological cooperation with the US and France. However, when the US demanded in return that India should refrain from building a gas pipeline from Iran via Pakistan, the government in

Delhi turned them down. India relies on the fact that the US has sufficient self-interests in accessing the South Asian market and is therefore not in a position to push through such a high price as the support of its Iran policy. And, as a matter of fact, Pakistan is an important ally of the US, in particular in the fight against terrorism. Pakistan, on the other hand, also wishes to leave other options open. Besides, it urgently need Iranian gas for its economic development.

The massive emergence of China and India as customers on the international energy market surprised even experts. However, the increase in oil and gas prices triggered differs from the oil crises of the 1970s and is of a structural nature. It will therefore continue to be an influence over the long term. At the same time, people tend to ignore the fact that two other East Asian countries, namely Japan and South Korea, have already been among the largest energy importers in the world for a long time. As members of the organization of industrialized states OECD and developed democracies, South Korea and Japan are important partners for Europe in the region.

Japan is the third largest economy and the third largest oil consumer in the world. It has almost no own fossil fuel reserves and is therefore highly dependent on imports or the development of alternative energy resources. Japan's strategy in the past few years has therefore been to diversify its imports, to increase its energy efficiency and to replace fossil fuel resources by renewable energy and nuclear power. Besides, Japan is obliged by the Kyoto climate protection protocol, signed in its own country, to reduce its greenhouse gas emissions further over the next few years.

The *New National Energy Strategy* published in 2006 by Japan's powerful Ministry of Economy, Trade and Industry (METI) sets a number of goals to reduce Japan's dependence on imports. For example, energy efficiency, expressed as one gross national product unit, is to be improved by 30 per cent between 2006 and 2030. The percentage of crude oil in primary energy consumption is to decrease from 50 per cent currently to 40 per cent by 2030. In the transport sector, the present 100 per cent dependence on oil is to be reduced to 80 per cent by the use of alternative fuels. At the same time, the percentage of nuclear energy and that of renewable energy is to increase. The core of the new national strategy, however, is increased engagement by Japanese energy companies in oil and gas production abroad, with assistance from the state. Japanese economists are nervously watching the aggressive advance of the Chinese on international energy markets. Japan's companies as well are to rely on financial and political support from their government when they go on their overseas shopping trips. Until now, Japan has obtained most of its oil imports from the Near East. As part of its diversification strategy, Japan's energy companies are currently investing in Russia, in the Caspian Sea and in Central Asia. The Japanese have set their sights on both Iran and Russia. The oil company INPEX is involved in the exploitation of the Iranian oilfield in Azadegan. The green light for this politically controversial business came directly from the office of former Prime Minister Koizumi, ignoring massive protests from the American Embassy against Japanese engagement in Iran.

What is strategically most important for Japan, however, is the energy partnership with its big neighbour Russia. The government in Tokyo has offered to contribute up to US$14 billion to a new oil pipeline from Siberia to the Pacific. In return, Japan demands that its companies should in future be allowed to obtain majority shares in Russian energy companies. The companies Mitsui and Mitsubishi are already involved in the exploitation of the largest oilfields in East Asia on the Russian island of Sakhalin in the Pacific. One obstacle to any closer political partnership between Japan and Russia at the moment, however, is the still unresolved territorial question regarding the Kuril Islands, which Russia occupied after World War II. Today, this dispute is not only about national pride, but also about how far Japan's continental shelf reaches into the northern Pacific and thus into potential offshore oil production territories.

Especially in this competition over the oil and gas supply from Russia, Japan is increasingly becoming a rival to its big neighbour China. This rivalry is evident in the year-long dispute over the route of a new oil pipeline from East Siberian production areas to the Pacific. At present, it is planned to lead the pipeline to the Russian harbour of Nakhodka in the Pacific, but also to build a branch pipeline to the Chinese refinery in Daqing. It would thus be possible to supply the markets in China as well as in Korea and Japan by pipeline and by tanker. There is also controversy between Japan and its neighbours in the region about oil production at sea. Over the long term, resource conflicts among East Asian Pacific rim nations can only be avoided if the position of maritime borders is regulated by international agreements.

In research, development and application, Japan competes with California and Germany over the first position in the utilization of photovoltaic solar energy. Japan is also a leader in the development of hydrogen technology. In spite of this, the country, where the atomic bombs of Hiroshima and Nagasaki were dropped, still focuses on extending nuclear energy, its unpopularity notwithstanding. In the past few years, there have been a number of serious incidents in Japan's reactors. Nevertheless, the fear of energy-political dependence has so far outdone all rational arguments against the use of nuclear power. Japan will only abandon the use of nuclear energy once it has been included in a regional energy supply system and no longer has to fear being cut off from its sea supply routes in the event of a political crisis in Russia or China. However, an obstacle to such closer regional cooperation is the fact that Japan's relations with its most important neighbours, China and South Korea, have deteriorated drastically over the past few years. The reason for this is Japan's past as occupying power during World War II, which has not yet been properly dealt with. But thanks to its enormous technological potential, Japan could actually become a pioneer in modern sustainable energy solutions for the entire region. Unlike neighbouring China and South Korea, Japan's energy consumption is only increasing slowly. Even though Japan's economy is six times as big as that of South Korea, it consumes only two and a half times as much energy. Japan is one of the leaders in energy-saving technology on the global market. If Japan wishes to assume the role of a regional leader, as it continues to affirm, one of its tasks should be to export energy-efficient technologies to its neighbouring countries as well, with help from its financially strong development bank. A sustainable development of energy utilization in the developing markets of East Asia would simultaneously be an important contribution to economic integration and stability in the region.

Like its neighbours China and Japan, South Korea depends almost completely on oil imports from the Near East. State-owned oil company Korean National Oil Corporation (KNOC) is therefore currently investing in Central Asia and in Azerbaijan to exploit alternative sources of supply. Nuclear power constitutes 40 per cent of South Korea's energy needs. The nuclear power stations are situated within range of North Korean missiles and are not sufficiently protected against military attacks. North Korea would therefore not need an atomic bomb to blackmail its neighbour to the south. In spite of this, South Korean environmental groups and nuclear critics have so far been unsuccessful in preventing further extension of the nuclear programme. Like the Japanese, the South Koreans fear that their energy-import dependent economy might collapse, should oil and gas imports be interrupted due to a political crisis.

Besides, South Korea, along with Mexico, is the OECD country with the lowest energy efficiency and the worst urban air pollution. On the Environmental Sustainability Index (ESI) that was published by the American universities of Columbia and Yale jointly with the World Economic Forum in 2005, South Korea ranks 122nd of the 146 countries that were compared, in between Liberia and Angola. By the way, North Korea ranks last.

Consequently, South Korea's energy policy is not sustainable for a number of reasons. South Korea is highly dependent on imports, its nuclear programme would be a theoretical target for military attacks from the north and pollution from industrial chimneys and exhausts is a danger to the health of its citizens. However, the new act on the promotion of renewable energies, drafted according to the German model, is a positive beginning. The explicit objective of this promotion is less dependence on energy imports.

China's new multinationals

While privately listed energy companies dominate the market and foreign companies play a role in Japan, South Korea and India, three of the largest oil companies in the world have recently been created in China with help from the state. In the country's development, China's new oil giants play a similar role as do Gazprom or Rosneft in Russia. They form part of the domestic modernization strategy of the Beijing government. In addition, they support China's active appearance on the world's resource markets.

China National Petroleum Corporation (CNPC) is China's largest national oil company. CNPC operates oil and gas projects everywhere in Central Asia, in the Middle East, but also in Burma, Sudan and in Latin America. CNPC is the leading international oil production company in Sudan. At present, oil from Sudan covers 10 per cent of Chinese demand. CNPC is also active in Nigeria, Algeria and Chad. One of CNPC's first international projects was its involvement in the Greater Nile Petroleum Operating Company in Sudan in the mid-1990s. Since then, China's state-owned oil companies have entered the African market in a big way. In this regard, the Chinese focus primarily on those countries in which Western companies have been less active so far. These are mostly countries struggling with considerable security and human rights problems, such as Sudan.

CNPC finalized its largest takeover abroad to date when it purchased the company PetroKazakhstan in October 2005. Jointly with Kazakh gas company KazMunaiGaz, CNPC is building a 1240km-long gas pipeline from the northwest of the Central Asian country to the province of Xinjiang in the eastern part of China. As a next step, the pipeline is to be extended to the Caspian Sea.

When state-owned Russian oil company Rosneft needed money to buy the oil production of the Yukos group, which has been broken up, CNPC granted it a loan of US$6 billion. Originally, it was Yukos that planned Russia's first private oil pipeline to China, jointly with CNPC. After company founder Chodorkovski had been arrested, the project was abandoned. Through their new relationship with Rosneft, however, the Chinese are still in business. CNPC dominates the market at home through its subsidiary Petrochina. Petrochina is responsible for oil production in the People's Republic of China and also holds the national pipeline monopoly. Petrochina has found oil in Xingjian, in China's Wild West, in Tibet and offshore in disputed coastal waters.

The second most important player in the business is the privately organized Sinopec Corporation. Even though Sinopec is formally organized as a private stock company, the Chinese state still holds more than 70 per cent of its shares. The remaining shares are held by Chinese banks and investors from Hong Kong. Under state-controlled Chinese economic policy, Sinopec is something of an experiment: a state-owned energy company is gradually opened up to private capital without relinquishing political control over the company.

State-owned China National Offshore Oil Company (CNOOC) was founded in 1982. Unlike CNPC and Sinopec, it is not predominantly focused on China's domestic market: jointly with foreign partners, it aims to exploit the oil and gas fields off the Chinese coast. A major oil and gas field under the sea was discovered in 2006, approximately 250km off the coast of Hong Kong. However, many of the oil and gas fields under the Chinese Sea that have been discovered in the past few years – and those with presumed future reserves – are situated in disputed waters. The borders of the respective territorial waters depend on under whose jurisdiction the many uninhabited islands and rock groups in the Western Pacific fall. China is fighting with almost all its neighbours, including Japan, the Philippines and Vietnam, over individual island groups, such as the Spratly Islands. The International Convention on the Law of the Sea does not provide a satisfactory answer to these unresolved territorial issues. In addition, Japan complains about the fact that CNOOC drills into and depletes joint oil and gas fields from the Chinese side.

Not least because the production offshore is burdened by these legal and diplomatic problems, CNOOC is also increasingly entering the international business. In summer 2005, CNOOC Ltd, the private arm of the Beijing state-owned company that is registered in Hong Kong, intended to buy Californian oil company Unocal at a price of US$18.5 billion. The majority of the shares in CNOOC Ltd are still held by the state, even though some shares have been sold to private Chinese entrepreneurs. The shock to American politicians and the energy industry when they realized that the Chinese can now actually bid in deals of this magnitude was so great that a storm of protest broke out, ending up in the US Congress. The latter threatened to pass a law to prevent Chinese engagement. CNOOC Ltd yielded to political pressure and eventually refrained from submitting a formal offer for Unocal. The Chinese then used the funds made available to buy their first offshore oilfield, close to Nigeria's coast, early in 2006 and have been doing business in Africa, as does CNPC. For the Chinese, the moral of this story was that they must try to improve their image in the US and in other Western countries in future. Thus, after the Unocal deal failed, CNOOC donated US$120,000 for the victims of Hurricane Katrina.

The leadership in Beijing knows that other countries watch China's rapid rise with unease. As a consequence, the doctrine of the 'peaceful rise' of China was developed. It states that the economic and political rise of the country should not be at the cost of others and that the new China is willing to become more engaged in international organizations such as the UN. The government in Beijing is very much interested in preventing the rest of the world from joining forces against a China that is perceived as a threat. They do not wish a repeat of the image loss that Russia and state-owned Gazprom suffered during the past few years. Similar to the Kremlin, the Chinese government tries to establish internationally competitive companies such as CNPC, CNOOC and Sinopec that – at least in the case of Sinopec – even formally fulfil Western standards of corporate governance. However, the objective of the Chinese energy giants is not only to make money, but also to serve China's economic and foreign policy.

China's 'peaceful rise' is increasingly being monitored by the international community. What will the foreign policy of domestically authoritarian China be like? How will the new superpower handle its relations with its neighbours? Are good corporate governance standards not worth the paper they are written on or will they have an effect in terms of a better protection of the environment and human rights?

China's new foreign policy

In April 2005, China's President Hu Jintao visited Saudi Arabia for the first time. Already in January of the same year, Saudi King Abdullah had visited China during his first foreign trip outside the Near East. Over the past few years, China has become an important customer for Near Eastern oil states and thus an important alternative to the Europeans and Americans. The title of an energy conference that took place in Dubai in January 2006 was 'Look East'. In 2005, Saudi Arabia alone supplied 17 per cent of Chinese oil imports. Along with the energy trade, other common interests are also growing. In Dubai, factories and shopping centres are built with Chinese capital. More and more students qualify at Saudi universities in subjects like Chinese language and history. On the other hand, Arabian businesspeople do not require a visa to travel to China, something they do need for entering the US and for which they have to wait months. At the moment, the US still guarantees safety in the Persian Gulf. However, the Chinese factor is becoming increasingly significant.

Along with its economic liberalization of the past few years, China also started reorientating its foreign policy. China's rise is accompanied by new frictions and conflicts of interests with its neighbours and other major economic powers. Economic success, China's growing role in global trade and the demands for political power at an international level cannot be analysed independently of each other. China's conduct in the UN Security Council, for example, in matters of human rights violations in oil-rich Sudan or Iran's nuclear programme, clearly show that economic considerations play a central role in the structuring of Chinese foreign policy.

At the time of President Hu's first visit to Washington in May 2005, the oil price passed US$70 per barrel for the first time. Even though the war in Iraq and the crisis over Iran's nuclear programme played their part and contributed to nervousness on the oil markets, analysts agree that it is China's emergence on the global market that will push prices up over the long term. China and the US have become the main competitors on the global oil market. Political and economic interests overlap in this regard. The US criticizes China's 'mercantilist approach'. By that, it refers to China's endeavours to secure the energy resources of other countries through long-term contracts, thus taking them out of the global market. At the same time, the US is not only the most important trading partner, but is also perceived as a major competitor, both in the East Asian region itself and in the creation of a multipolar world order. Since World War II, the US functions as the police force in the East Pacific and has stationed large contingents of troops both in Japan and in South Korea. The US Navy controls the sea routes in the Pacific and in the Indian Ocean and thus shipping and tanker routes central to the Chinese economy.

In East Asia, China competes against Japan and South Korea. Relations with Japan not only reflect the competition over access to Russia's energy resources, but also the territorial conflicts over presumed oil and gas resources underneath the disputed offshore waters between the two countries. In the meantime, China has replaced the US as Japan's most important trading partner. On the other hand, mutual dependences in the region have increased so much over the past few years that think tanks and planning experts in Tokyo and Beijing are discussing closer political and economic integration. The goal would be to ease political tensions and to build a common economic region. This could be done following the example of the onset of European integration in the 1950s.

China's cooperation with South East Asian countries and with India is also stronger than it has ever been. By 2010, a free-trade zone is to be created with the economic organization Association of South East Asian Nations (ASEAN). China imports oil from Indonesia and from Myanmar, the latter being governed by a military regime. Oil from Myanmar – one of the poorest countries in the world – can be imported over land. The sea route via the Strait of Malacca can thus be avoided.

Closer rapprochement with India has begun since the US started working closely with Pakistan, India's regional adversary. Energy-political interests, for example, access to Iranian oil and gas, play an important role in this partnership of convenience as well. In 2004, China and India signed an agreement to prevent their companies getting in each other's way when exploiting natural resources in other countries. In the nuclear field, however, China continues working together with Pakistan.

Chinese national companies are trying to secure exploitation contracts for themselves even beyond the Asian continent, for example, in Africa and Latin America. The first Sino-African Summit took place in Beijing at the end of 2006. Concurrent with its energy-political engagement, Beijing also intends to create a stronger presence in Africa as far as development policies are concerned. At national level, a system of strategic oil reserves is to be established in the next few years similar to the one created by Western industrialized countries after the oil crisis in the 1970s.

China's policy in Central Asia

Central Asia has a key position in China's consideration of the long-term security of its energy supply. Politically and diplomatically, the Chinese rely on regional integration with the five Central Asian republics of Kazakhstan, Uzbekistan, Turkmenistan, Tajikistan and Kyrgyzstan. The relationship with Russia, which had always dominated and controlled Central Asia since the times of the tsars, fluctuates between cooperation and competition. Both powers treat Central Asia as their own regional backyard. The eventual arrangement between Moscow and Beijing will have a lasting influence on the political and economic development opportunities of the Central Asian states. An expression of such cooperation is the Shanghai Cooperation Organization (SCO), which was founded jointly by China and Russia. Nevertheless, China and Russia remain economic opponents in the region. The US is also economically, politically and militarily present, although not as dominant as in the Near East. Other important powers with interests in the region are India, Iran and Turkey.

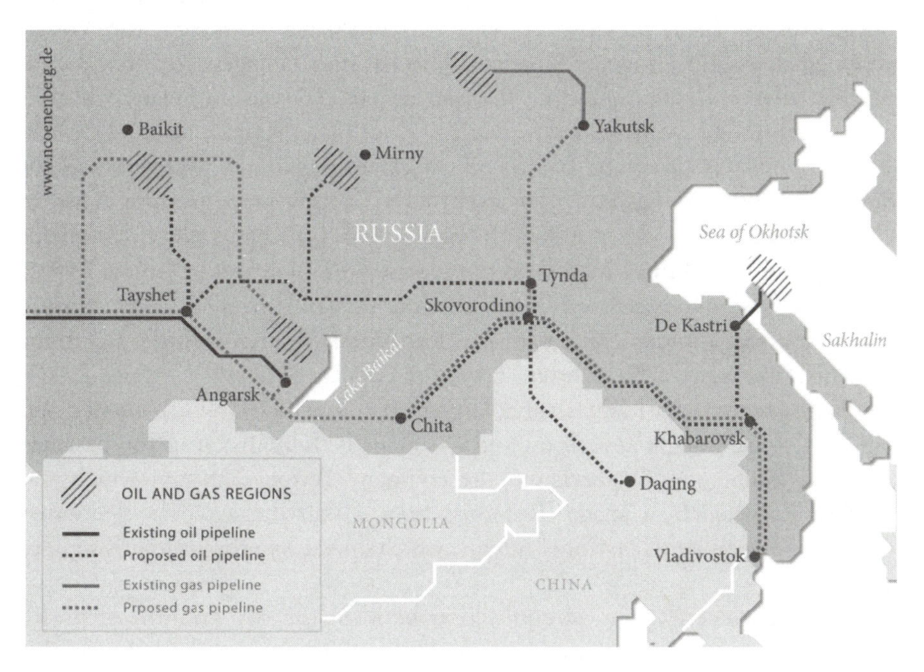

Figure 4.1 Oil and gas pipelines in China and Central Asia

The EU, on the other hand, has never really gained a foothold in Central Asia since the end of the Cold War. From a political perspective, this is a serious lapse, not least because of the energy-political interests of Europe in the region. The

German Federal Government is therefore working on a Central Asia strategy for the EU, which is to bring the economic and political interests in the region into line with the progress of democracy and human rights. After all, Germany is the only EU member state to have its own embassy in each of the five Central Asian republics.

Currently, Europe obtains natural gas supplies from Turkmenistan via Russian pipelines. New pipelines are being constructed that will link the entire region around the Caspian Sea to the Western oil and gas trade network. The economies of the Central Asian nations are not diversified and rely exclusively on the production and export of its energy resources. Their governments are authoritarian. Political opposition, which often comes from ethnic minorities and radical Islamists, is violently suppressed. Over the long term, all Central Asian states should therefore be regarded as unstable. This instability may even spill over to Europe's immediate neighbourhood; to the Caucasus region, for instance.

Because of its gigantic oil resources, Kazakhstan is often referred to as the new Kuwait. According to estimates by the IEA, the country has 3.3 per cent of the world's oil resources and 1.7 per cent of the gas fields. This is equivalent to the size of the fields in Libya or Nigeria. Until a few years ago, Kazakhstan exported all its oil via Russian pipelines. This has since changed: in 2003, the gas pipeline of the Caspian Pipeline Consortium (CPC) was completed. It runs underneath the Caspian Sea to Azerbaijan. From there, Kazakh oil can be transported by tankers or via the Baku-Tbilisi-Ceyhan pipeline to Western markets. Since July 2006, oil has also been exported via a new pipeline from Atasu in Central Kazakhstan to Dushanzi in the province of Xinjiang in eastern China for the first time. Until then, road tankers were used for oil export to China. Export quantities were thus restricted. A gas pipeline that runs parallel to the existing one is currently under construction. From Atasu, both pipelines are to be extended into western Kazakhstan, up to the Caspian Sea.

At the same time, China's three largest oil companies CNPC, Sinopec and CNOOC have bought several oil and gas fields in Kazakhstan in the past few years. In 2005, CNPC purchased the company Petrokazakhstan, which was previously owned by Canada. Rumours have it that the takeover of another Canadian oil company, National Energy, which operates in the region, is soon to be finalized.

The close energy-political connection between the two countries has two important consequences. China provides military support to the authoritarian Kazakh government under President Nazarbayev by supplying weapons to the Kazakh army. Under the guise of the fight against terrorism, the two countries cooperate in the suppression of ethnic and religious opposition movements. The province of Xinjiang in western China, which borders on Kazakhstan, is gaining in strategic significance for China as an oil and gas transit corridor. Opposition groups and free media of Muslim population groups in Xinjiang are therefore systematically suppressed. Instead, the leadership in Beijing settles representatives of the ethnic majority, the Han Chinese, in its eastern province, as they did in Tibet. Today, Han Chinese hold key positions in Xinjiang's politics and economy.

Like Kazakhstan, the other Central Asian republics are also trying to escape the political and economic grip of Russia. This dependence is primarily evident from the fact that, until the end of the 1990s, all Central Asian countries had access to the global market only via the pipeline network stemming from Soviet times. To free themselves of this dependence, all states in Central Asia are looking for foreign investors and political alternatives to the power of the Kremlin. China is a welcome partner in this regard.

Back in 1997, when Gazprom stopped the transport of Turkmen natural gas via its network to enforce higher transit prices, Turkmenistan painfully experienced its one-sided dependence, when its national income was reduced dramatically. Due to the budget crisis, the government came under pressure at home and soon gave in to the Russians. Today, Turkmenistan continues selling its gas to Russia at prices far below the global market price. Cheap Turkmenistan gas also subsidizes the prices at which Gazprom sells gas to Belarus and other allies of the Kremlin, subsidies applied for political reasons. As hardly any Western company wishes to invest in Turkmenistan, it has now turned to China. In April 2006, Turkmen late head of state Saparmurat Niyasov and China's President Hu Jintao signed a 30-year agreement on the export of gas to China and on the construction of a 4000km-long pipeline from the Amudarya gas field to Shanghai. This pipeline is planned to be completed by 2009. However, the industry magazine, *Energy Economist*, has warned that Turkmenistan will have difficulties in fulfilling its contract with China in addition to its existing supply contracts with Russia, Ukraine and other Eastern European countries. There is simply not enough gas available in the current production fields. Should Turkmenistan no longer be able to comply with the delivery quantities it has promised, a decision will be made based on which of its customers has gained the greatest political influence in the region by then. The race between the Russians and the Chinese is not over yet. Other customers, for example, the Eastern European countries, which have so far benefited from cheap Turkmenistan gas, could then be left in the lurch – another reason for the Eastern Europeans to move away from their dependence on cheap gas imports.

Apart from Kazakhstan and Turkmenistan, China's main focus is Uzbekistan. Uzbekistan is the eighth largest natural gas producer in the world. Its oilfields are also significant. In 2005, two of the largest Chinese energy companies, CNPC and Sinopec, established extensive joint ventures for the production of oil and gas together with national Uzbek production companies. Here again, the Achilles' heel in China's energy strategy is its lack of a pipeline network. To date, all Uzbekistan's natural gas exports use the pipeline network of Russian monopolist Gazprom.

Apart from its resources, Central Asia is also important to China as a transit region to obtain access to the oil and gas fields in the Caucasus and in Iran, and to overcome its unilateral dependence on oil imports from the Middle East. The oil that China imports from there has, until now, exclusively been transported by sea. However, the sea routes through the Persian Gulf, through the Strait of Malacca and past Taiwan are controlled by the US navy. China's energy security has thus been guaranteed by US military power. Over the long term, the Chinese will wish to escape this dependence through their involvement in Central Asia.

China's Africa policy

China's oil diplomacy in Africa has become competition for former colonialist powers France and the UK as well as the US. In socialist states formerly under Soviet influence, such as Angola, China has replaced the Russians as the most important economic and political partner. In November 2006, 41 African heads of state and government attended the largest Sino-African Summit to date held in Beijing, where China promised an increase in its development aid, low-interest loans, export guarantees and customs duties relief. Investment is primarily planned in Africa's natural resource sector.

Today, 28 per cent of Chinese oil imports come from Sudan, Angola, Congo, Nigeria and other African countries. Africa is already contributing 12 per cent to global oil production – a percentage on the increase. While production is going down in other regions in the world, the energy industry reckons that the production from African wells will increase by another 30 per cent between 2006 and 2010. War, political instability, corruption and organized crime are still the order of the day in many African states – including those with considerable oil and gas reserves. However, the international oil companies' willingness to take risks has increased in times of high prices. Even though the US and the two former colonialist powers France and the UK still contribute 70 per cent of foreign investment in Africa, the Chinese are catching up. Their advantage is that they also invest in politically problematic countries and regions. Besides, Chinese state-owned companies do not think in terms of short-term profit, but rather of long-term securing of resources.

For instance, China has become the largest oil customer of Angola following the end of a 30-year long civil war in that country. In return for access to the Angolan market, China provides generous loans. The government in Luanda could thus afford to refuse a competing offer from the International Monetary Fund (IMF) that would have been conditional on economic reforms.

Another good business partner of China is the Islamic regime in Sudan. While Western countries boycott Sudan because of serious human rights violations in the south of the country and in Darfur, Chinese oil purchasers are not that particular. When challenged by the *New York Times*, Chinese deputy foreign minister said: 'Business is business. We try to separate business from politics.' Obviously, this is not completely true. China is also interested in a peace agreement in Sudan. Even if one is not interested in democracy and human rights per se, it is impossible to establish a stable business environment under the conditions of civil war and the constant danger of terrorist attacks. Chinese companies' oil business in Africa is therefore supported politically by the government in Beijing. For instance, China sold patrol boats and other weapons to Nigerian security forces. These can be used to defend the oil production rigs in the Niger delta against local rebels and organized crime. The US government turned down a similar request from the Nigerians because of the suspicion of corruption among Nigeria's military and police forces.

In Africa, not only are business interests of the old superpowers of the West and self-confident new superpower China in conflict, but they also have different philosophies on what rules should apply in international business. Even though the Europeans' and Americans' complaint that China was doing business with little respected governments such as Sudan might appear hypocritical, the world does need a code of conduct for international investment and the activities of multinational groups over the long term. After all, there are no NGOs in China to criticize the environmental sins and human rights violations of Chinese companies abroad, and to enforce remedies for any such transgressions.

Nuclear competition

The major powers in Asia have become involved in a dangerous nuclear competition. In 2005, China had an estimated 400 nuclear devices, while India and Pakistan were believed to have 50 each. In 2006, North Korea tested its first atomic bomb. Other countries, such as Japan and South Korea, have the technical capacity to produce nuclear weapons within only a few months. Not to forget Russia and the US that have stationed tactical nuclear weapons in South Korea and on aircraft carriers along Asia's coasts. Everywhere in the region, it is evident that military nuclear programmes were developed under the veil of civil programmes. Iran is the most important example in this regard.

The US, Russia and France stand to make a lot of money from the export of nuclear technology to the nuclear growth market of Asia. With the Indo-American Nuclear Agreement, the US is trying to build up India as a regional counterweight to China. The Russian government has offered Iran cooperation in nuclear technology, at the same time aiming to keep the Iranian nuclear weapon programme under control.

So far, the states in the region, instead of trying to prevent a nuclear armament race, are rather promoting it. North Korea and Pakistan have closely cooperated in the development of their military nuclear programmes. India's nuclear weapon programme was subsequently legitimized when the Indo-American Nuclear Agreement was signed in 2006. Apart from the Iranian nuclear programme, the international community is currently mainly worried about Pakistan's 'Islamic bomb'. Western governments hope that the military dictatorship in Islamabad, which already has nuclear capabilities, will remain as stable as possible to prevent it from falling into the hands of Islamic terrorists.

The counter to regional competition over nuclear energy and weapons is once again stronger regional and international cooperation. One example in this regard is North Korea. In the mid-1990s, the US government, then under President Clinton, negotiated an agreement with the North Korean regime which provided for energy supplies from China and the US to North Korea and, as a quid pro quo, the discontinuation of Pyongyang's military nuclear programme. China continues to deliver hard coal to North Korea. However, the US soon stopped their oil deliveries due to domestic disputes over the orientation of its North Korea policy, which in turn resulted in North Korea reassuming its nuclear weapon programme. In spite of this, the only way to reach an agreement is for North Korea to abandon its controversial nuclear weapon programme in return for economic aid and energy supplies from its neighbours.

Regional cooperation, regional competition

China's energy hunger not only affects regional, but also global security and stability. Its search for oil and gas along its coasts led to territorial conflicts with its neighbours Japan, Vietnam and the Philippines. An alliance with Russia is to secure delivery by land, while delivery by sea is to be secured by reclaiming Taiwan. At all these places, China confronts American influence zones. In return for close economic cooperation, China supports Central Asian dictatorships, such as Uzbekistan and gas-rich Turkmenistan.

China's growing economic influence is accompanied by diplomatic initiatives that China employs to position itself as a regional counterweight to the US. China is also negotiating bilateral trade agreements with resource-rich states in Africa and Latin America. In many cases, the essence of the trade is oil for weapons.

China and India have not only the purchasing power, but also the investment funds for extending and modernizing the infrastructure in Russia and the former Soviet republics in Central Asia. In return, Russia could supply the Asian economic region with natural resources and energy. Already today, Chinese state-owned companies invest heavily in oil- and gas-exporting countries Kazakhstan and Turkmenistan. China is also pushing for the construction of additional supply pipelines, above all from Russia's far east. However, it faces strong competition from South Korean and Japanese companies.

On 15 June 2001, China and Russia founded the Shanghai Cooperation Organisation (SCO) jointly with the four Central Asian countries Tajikistan, Uzbekistan, Kyrgyzstan and Kazakhstan. The original task of the organization was to resolve border disputes between the former Soviet republics and China. This challenge was met successfully, so that the alliance could turn to other tasks. Since then, the six countries have cooperated politically, economically and militarily via the SCO. Russia and China supported Uzbekistan's dictator Karimov after the latter put a bloody end to a protest march in the city of Andijan. The first joint military manoeuvre of the SCO took place in 2007. A new oil pipeline connects Russia, Kazakhstan and China. When the US, after the terror attacks of 11 September 2001, wanted to extend its influence in the region and to build military bases everywhere in Central Asia, the states of the SCO tried to restrict and later push back American influence. The US military bases in Uzbekistan have been shut down in the meantime. Kyrgyzstan, on the other hand, wishes to continue earning money from US military presence. Kazakhstan, in turn, does not trust its allies and tries to find a balance between the three major powers Russia, China and the US. For example, Kazakhstan not only participates in military manoeuvres of the SCO, but also took part in the NATO exercise 'Steppe Eagle' which took place on Kazakh territory in September 2006. With the new Transcaspian pipeline, which was planned on the initiative of the US, gas is to be supplied from Kazakhstan and the Caspian basin to the West, bypassing Russian territory. The new power-political centre of gravity that has been created with the SCO in the middle of Asia is also attractive to other regional powers. Pakistan, Iran, India and Mongolia have already applied for observer status. If India and Pakistan were to join the SCO, four of the eight nuclear powers of the world would sit at one table. Even beyond the borders of the region, the SCO has been pursuing an active alliance policy, for example, with the Gulf Cooperation Council (GCC) of the Arab states. Here again, the common interest of both these groupings is to prevent dependence on the US.

Following in the footsteps of Marco Polo – is the EU looking eastwards?

Marco Polo was the first European to travel to China. Today, thousands of businesspeople visit the Heavenly Empire every day. While economic relations between Europe and China are close, this holds less true for the political partnership.

The objective formulated in the European Security Strategy (ESS) is to develop a 'strategic partnership' with China – and also with India to which much of what is referred below also applies. Even though the term 'strategic partnership' is purposely imprecise, it can still be said that it refers to a comprehensive, long-term relationship to the benefit of all participating parties. Naturally, of paramount importance are the political and economic objectives that the relationship with the important partner China is to serve and on what values it is based. In the past few years, China has become an increasingly important player in foreign and security policies for the West. This state of affairs is mainly due to China's growing economic strength and its increasingly active role – both in regional committees and multilateral systems. Without the cooperative integration and active involvement of China, numerous challenges concerning security policies can no longer be resolved. Today, the question is how China uses its new role and stature. The dispute over Iran's nuclear programme is an example in this regard. The Chinese leadership is striving for stronger engagement within the UN and other multilateral organizations as well as a more active role in the resolution of regional security conflicts, true to the doctrine of the 'peaceful rise' of China. At this stage, however, Sino-European relations are still in a 'pre-political' stage, so to speak. They are dominated by the economic and commercial interests of the large member states. There is generally a lack of strategic considerations on the future role of China in the multilateral system and within the Euro-Asian security structure.

Actually, the EU is the partner of choice in China's foreign policy. Europe is situated far enough away not to constitute a direct threat to China, but also carries the necessary economic weight to act as a potential counterweight to Russia and the US. As far as energy imports are concerned, both China and the EU depend on Russia and have a vital interest in not playing one off against the other. Thus, China and the EU signed a joint declaration on energy security at their summit in September 2006. Both sides wish to cooperate closely on technological policies, the promotion of energy efficiency, and the creation of common markets. Besides, both partners wish to better coordinate their client behaviour vis-à-vis energy exporter Russia. Even if China has been the focus of European politics in the last few years, the EU should not forget the democracies in India, Japan and South Korea. These are the ideal partners for Europe to resolve common challenges within the framework of multilateral cooperation.

It will be interesting to see how the development in economic and foreign policy of democratic India will differ from that of authoritarian regime China. In spring 2006, India signed a far-reaching energy-political cooperation agreement with the US, which, although not yet ratified by the parliament in Delhi, also is to serve as blueprint for similar agreements with other Western partners. At the same time, cooperation in the area of energy efficiency and the promotion of renewable energies is to be strengthened. In return, India has, if not cancelled, then at least postponed indefinitely the construction of a gas pipeline from Iran. Even though the two sides do not say so openly, Indo-American cooperation aims at balancing China's growing influence in Asia. If the US tries to achieve a balance of power in Asia with its policy today, it should be in Europe's interest to encourage regional economic and political cooperation according to the EU model. The core of crystallization in this regard could be ASEAN, the East Asian economic alliance. Also in the area of clean air, East and South Asian countries have begun transnational cooperation, following the model of the agreements that Western and Eastern European countries reached at the time of the Cold War to fight against transnational air pollution and acid rain.

Unlike Europe, where the confrontation among European national states after World War II was resolved through economic and political integration, certain historic events of war, occupation and eviction in East and South Asia still need to be dealt with. This difficult step will be the prerequisite for solving the common problems of the region in a cooperative alliance.

Chapter 5

A Common European Energy Policy

Even though the EU is the greatest economic and trading power in the world, so far it lacks its own energy policy. Consequently, there can be no coherent EU foreign policy, as the EU will remain vulnerable to blackmail in this crucial question of economic survival. As Europe will remain dependent on energy imports from its southern and eastern neighbours for the foreseeable future, the common energy policy must not be restricted to the EU in the narrower sense, but must also include the neighbouring countries in Eastern Europe and the Near and Middle East. Key countries and regions in this regard are Ukraine, Turkey and the states of the Southern Caucasus. They serve as a political bridge and as transit countries for energy imports into the EU. Therefore, these countries must be brought closer to the EU politically and economically. However, European energy foreign policy focuses mainly on Russia. This neighbourly relation, which is sometimes conflict-ridden, should develop into a partnership based on cooperation and international agreements for jointly shaping the European continent economically and politically. Since the beginning of the 1990s, the EU has been striving to sign an agreement – the European Energy Charter – on energy-political principles with all its neighbours in order to combine the economic, environmental and security interests of all parties involved into one integrated concept, and to regulate it by international treaty.

The declared objectives of the EU not only include a reliable and affordable, but also an environmentally friendly, energy supply. The development of environmentally friendly alternatives could also reduce the dependence on imports of fossil fuels, thus increasing the EU's energy security. Apart from creating the physical infrastructure for Europe's energy supply, the main goals of EU policy are therefore climate protection and securing the stability and good neighbourly relations on the European continent.

Europe's energy dependence

The European Security Strategy of December 2003 reads as follows: 'Energy dependence is a special concern for Europe. Europe is the world's largest importer of oil and gas. Imports account for about 50% of energy consumption today. This will rise to 70% in 2030'. A new partnership and cooperation agreement between the EU and Russia should also embrace those principles (European Council, 2003). Energy imports come mostly from regions with fragile security and political situations, such as the Gulf States, North Africa, Russia and Central Asia.

If the trend forecast by the European Commission continues, the EU's dependence on energy imports will keep rising in the next few decades. First and foremost, this is true for imports of natural gas, as the trend to change power generation from coal to environmentally friendly gas will continue. The European Commission estimates that natural gas consumption of the EU members will increase two-and-a-half fold between 2000 and 2020. In Germany, the percentage of power generation from gas will increase from 17 to 25 per cent by 2020, in the UK possibly from 25 to 40 per cent. Domestic resources, such as oil and gas from the North Sea, on the other hand, are running out. Russia is already the most important exporter of fossil fuels and of natural uranium to the EU. The Russians are therefore the most important alternative to the crisis-prone export countries of the Near East and are taking away from them their hegemonic position in the petroleum and gas market. It is not by coincidence that the construction of the first transcontinental gas pipeline between the former Soviet Union and Germany was started after the first oil embargo of the Arab states in the mid-1970s. According to the Commission's forecast, by 2020, oil imports from Russia will increase by 20 per cent and gas imports by 150 per cent, over 2000 levels. Aside from Russia, Iran has the potential to become the EU's most important gas supplier. Along with Ukraine, Turkey would thus become the most important transit country for Europe's energy supply. Imports from North Africa will also increase. In future, liquefied gas will also be imported by tanker from countries such as Nigeria and Venezuela.

If the EU wishes to secure its natural gas supply beyond the year 2020, it faces three geostrategic challenges: how to structure the relations with Russia; will Iran successfully be integrated into the international political system and global economy; and is Turkey to become a member of the EU?

If, at the same time, the EU wishes to reduce its economic and political dependence on energy imports from surrounding countries, a fundamental shift in European energy policy away from the dependence on oil and gas to a more efficient utilization and increased use of renewable energy would be required. Any security problems resulting from economic dependence on neighbours such as Russia and Iran would be, if not immediately and completely eliminated, then at least reduced considerably.

Historic roots

European unification began with energy policy. The European Coal and Steel Community (ECSC) was founded by the Treaty of Paris on 18 April 1951. It created a common customs union for those two commodities, which were essential for Europe's reconstruction after the war. The ECSC was based on the so-called Schumann Plan of the then French foreign minister. On recommendation of the French pro-European politician Jean Monnet, Robert Schumann wished to create common interests and stability in Europe by collectivizing coal and steel, formerly important goods during the war. This so-called 'Monnet method' of creating common political interests and a culture of political cooperation through economic integration has since then been established as the principle of European politics. EU foreign policy also follows the Monnet method and focuses on economic cooperation and political integration.

On 25 March 1957, the same date on which the European Economic Community was established, which would later develop into the EU, the European Atomic Energy Community (EURATOM) was launched. The task of EURATOM is to promote nuclear technology in all member states, to share nuclear-technological know-how, and to control the production of fissionable materials jointly. The ECSC treaty was later integrated into the EC treaty. The EURATOM agreement is still in force unchanged, and is to become a part of the Treaty of Lisbon, the scaled-down version of a European Constitution. In spite of the fact that European integration started with the joint management of coal and the joint promotion of nuclear energy, these beginnings have failed to produce an integrated concept for European energy policy. This has been impossible as the interests of important member states were, and continue to be, too diverse.

Resource nationalism and national champions

Until the partial liberalization of the energy markets in the 1990s, most European energy suppliers had been controlled by the various individual states and used as part of their national industrial and special interest-driven policies. To this day, state-owned French electricity company Electricité de France (EdF) focuses on nuclear energy. In Germany, on the other hand, climate protection and the promotion of renewable energy are playing an increasingly important role. While Poland generates most of its electricity and heating power from hard coal, the Czech Republic uses lignite, whereas the Baltic states and Slovakia depend almost 100 per cent on gas imports from Russia, a dependence they would like to be free from. Germany and the Netherlands, on the other hand, do excellent business with Russia's energy industry. In the meantime, Russian Gazprom has also entered the German gas distribution network buying interests in municipal power companies. The UK, on the other hand, has fended off attempts by the Russians to buy British gas company Centrica. So far, the British can afford this policy. Most of their natural gas comes from Norway and North Africa. However, as North Sea oil and gas are running out, the UK's import dependence will probably increase to 90 per cent by 2020. Then the cards will be reshuffled, presumably also with Russian involvement. The situation of Southern European EU member states, such as Italy and Spain, and in part France, is different. They purchase their oil and gas imports from North Africa and the Near East. Therefore, they do not look eastwards, but rather to the southern neighbours of the EU in the Mediterranean.

After energy markets were liberalized in the 1990s, Europe's major energy companies have gone on a global shopping trip. Since previously existing territorial monopolies were gradually dismantled within the EU, Europe's energy companies – regardless of whether they are still state-owned or have been privatized – must, for the first time, hold their ground against the competition. The governments in Germany, France and Italy are therefore transforming their formerly state-owned companies into competitive national champions. The strategy can be called forward defence and comprises two elements that complement each other.

At home, undesired competition from abroad is prevented by legal tricks and dodges. In Germany, former state-owned companies have virtually maintained their territorial monopolies. Small suppliers, such as Dutch power supplier Nuon or the green energy company Lichtblick, have only recently begun trying to establish a niche as a low-priced supplier to private electricity consumers. Only Swedish Vattenfall group has succeeded in buying into the east of Germany where territorial monopolies had not been that stable. To gain a foothold in the lucrative German market, Vattenfall has made numerous political concessions. For example, Sweden undertook to continue lignite power generation in the Lausitz region of East Germany. From an economic point of view, the construction of modern gas power plants would have been more profitable. In Berlin, Vattenfall even took over the sponsorship of ailing football club Hertha BSC from former municipal electricity supplier BEWAG.

While it is hard to play against the top dogs such as E.ON and RWE in Germany, it is almost impossible in France: EdF, still run by the state, owns the entire power network in the country. For this reason, foreign suppliers only have a chance in the border regions.

Through the captive market on home ground the new national champions can keep prices high and earn a secure income, which is then used for purchasing international companies. This is the reason why there has been a wave of mergers in wealthy Western Europe, during which a few big players have swallowed up the remaining small ones. The growth market lies in Eastern Europe, where so far only the former state-owned energy company of the Czech Republic and the Polish oil company Orlen have been able to hold their ground against the Western European giants in the industry, even going on a shopping trip to other Eastern European countries such as Bulgaria and Lithuania. Orlen has actually got into major trouble by buying a refinery on the Lithuanian coast that was also eyed by Russia's gas monopolist Gazprom. Over the medium term, however, the markets of developing countries in Africa, Asia and Latin America hold the greatest potential. Spanish companies are at the forefront of those investing in the emerging markets of Latin America. There, however, the Europeans will be facing stiff competition from the US and from emerging economies such as China, India, Brazil and Russia themselves.

On the crest of the ongoing European merger wave are the two German industry giants E.ON and RWE. Following a number of mergers, four major electricity companies have been created, namely E.ON, RWE, Vattenfall and ENBW, which are also very active in the gas sector. However, thanks to German legal regulations on energy management and supply, public utility companies and independent operators, for example, wind generated energy suppliers, also have a chance of survival on the domestic market. The big four, on the other hand, have expanded throughout Europe in the meantime, buying companies in Eastern Europe in particular. RWE subsidiary Energy Czech follows the example of its parent company and is heavily buying into Slovakia, Romania and Bulgaria. The new Polish government wishes to do as the Czechs do: Poland's national energy company, which was partially privatized in the 1990s, is to be brought back under government control and developed into the national champion.

Apart from the German energy giants, the European market is dominated by French EDF, French gas and water company Suez, and the Italian Enel group. This is not a peaceful coexistence, but rather a fierce struggle for market share and survival. For example, Italy's Enel was to take over the French Suez group in 2006. The French government attempted to fend off this takeover by merging private company Suez with state-owned Gaz de France. This protectionist manoeuvre caused considerable political disgruntlement between the governments of France and Italy.

In a dramatic fight for control over the Spanish gas company Endesa, E.ON offered a price of €50 billion. The Spanish government, however, used all legal tricks to ward off this takeover. Initially, the Spanish wanted to merge Endesa with its local competitor Gas Natural, following the French example. When E.ON submitted the better offer, the merger was only authorized subject on condition that E.ON sell 30 per cent of the power plant capacity to competitors. E.ON itself took over competitor Ruhrgas, against the recommendation of the German Monopolies and Mergers Commission, but with special permission from the Federal Minister of the Economy. Finally, a significant share of Endesa was sold to Italy's Enel, but only after Italy's and Spain's prime ministers met to arrange such a deal and fend off the German competitor. Some years ago E.ON itself had bought competitor Ruhrgas against the explicit recommendation of Germany's Anti-trust Agency. Then-economic minister Müller had to give special permission. According to the EU competition protection regulations in force, the creation of a national monopoly of this size would not have been authorized. In order to prevent these protectionist tendencies from undermining the domestic energy market, the European Commission is trying to enforce EU competition protection regulations against decisions of national governments – however, it does not always succeed.

The European Commission has therefore endeavoured over an extended period of time to obtain a mandate from Europe's governments to plan a common energy policy and to represent the EU in dealings with third parties. Because of different economic alliances and the resulting national energy strategies, Europe appears split when dealing with export nations, such as Russia and the Near East, monopolies like OPEC and competitors for scarce resources, such as China and the US. The European member states are therefore eventually played off against each other. Their overall interest falls by the wayside. For this reason, a joint European energy policy could become one of the key projects of further integration of Europe into an economic and political union.

The new EU member states in Central Europe, such as Poland, which are especially dependent on energy imports – primarily from Russia – are starting to realize that only a common European energy strategy can ensure diversification of supply and restore autonomy in energy policies. First steps in this regard have already been taken. Poland, for instance, is planning a pipeline jointly with Ukraine to transport oil from the Caspian Sea to the port of Gdansk on the Baltic Sea. Close to the old Hanseatic city, construction is under way of a new liquefied gas terminal. The gas would come from Norway, while at least part of the investment funds come from the EU budget. Besides, coal-rich Poland needs assistance from Brussels for restructuring its uneconomical coal-mining industry and for its fight against air pollution by building new power plants. Poland will only be able to fulfil the obligations on climate protection that it accepted as part of the EU if it focuses on renewable energy, efficiency and also on highly efficient CO_2-free new carbon technology.

In Poland, as in the other Central and Eastern European countries, the restructuring of the energy and agricultural sectors could go hand in hand. Poland needs domestic and environmentally friendly energy. Polish agriculture requires new markets. Electricity and fuel generation from biomass could bring the country closer to achieving both objectives. As a corollary, the EU could reduce its agricultural surplus and gain new breathing space in negotiations at the WTO. However, this would require part of the immense EU agricultural subsidies to be redirected to biomass production.

The best example of how things should not be done is the planned North Stream gas pipeline – or Baltic Sea Pipeline, as it is commonly referred to – that is to be used in future to transport natural gas from Russia to Western Europe and thus above all to Germany. State-owned Russian Gazprom has a 51 per cent share in the construction consortium, with E.ON subsidiary Ruhrgas and BASF subsidiary Wintershall holding 24.5 per cent each. The pipeline is to run for

1200km underneath the Baltic Sea and, as from 2010, is to transport 27.5 billion m^3 of natural gas annually from Siberia to Germany and Western Europe. The contract for the construction of the Baltic Sea Pipeline was negotiated with close involvement of the German and Russian governments. Former German Chancellor Gerhard Schröder has since been appointed as Chairman of the Board of the pipeline consortium. The project is politically explosive because Germany arranges its energy supply with Russia without any participation by its EU partners. Germany wishes to become the hub for energy imports from Russia to Western Europe. This policy was initiated by the Schröder government and is continued by the current federal government of Angela Merkel. The German government hopes to ensure privileged access for German companies to the Russian energy wealth and a key political position in the relationship between Russia and the EU. This, of course, is to the detriment of European integration. Poland and the Baltic countries are particularly unhappy about the fact that the Germans have cut them out. Branch pipelines could actually ensure a better future supply of natural gas to these countries as well. In the end, the Baltic Sea pipeline could fail because the construction over the fragile bed of the ocean causes too many problems for the environment. Lithuanian and Swedish scientists have warned for a long time already that thousands of barrels containing poisonous gas from World War I are still lying at the bottom of the Baltic Sea. Nobody knows where these barrels are located and what could happen if construction equipment were to destroy them.

Even if the EU were to succeed in better coordinating the energy policy of its member states, or even planning it at European level in the foreseeable future, it must be realized that the political arena where decisions on the exploration, transportation and utilization of energy are to be made extends far beyond its borders. In future, oil and gas will come from Russia, Central Asia and the Mediterranean. The transportation comes through countries in close proximity to the EU. The states of East Asia at the other end of the Eurasian landmass are rapidly becoming competitors, requiring that a cooperative relationship be established with them.

The European Energy Charter

In December 1994, the European Energy Charter – a basic treaty for a common pan-European energy policy – was signed. The crucial objectives of the Energy Charter are the creation of a regime of energy management based on the principles of market economy, the regulation and promotion of trade and investment in the energy area, and the mediation in any disputes that might arise from these challenging undertakings. From the very beginning, it was clear that a balance would have to be found between the different interests of energy exporters, importers and transit countries when negotiating such a new type of fundamental agreement for an energy-political cooperation for all Europe. Besides, investment and business interests of the private energy industry have to be protected while also accepting the fact that the energy sector is still state-controlled or parastatal in many countries. Lastly, apart from the interests of the economy and the investors, the other objectives of EU energy policy should not be forgotten; these primarily include the protection of the environment and climate.

All states of Europe, as well as Russia, Japan and Australia are members of the European Energy Charter. The US and Canada as well as some Mediterranean rim states enjoy observer status. In a first version of the Charter formulated in the beginning of the 1990s, the objective of 'creating an energy community of the countries on both sides of the Iron Curtain based on the complementarity of Western markets, the capital, technology and the natural resources of the Eastern states' was evoked in an almost poetic manner.

As ambitious as the basic objectives of the European Energy Charter are, it has a decisive shortcoming. The Charter has not yet been ratified by Russia, the country whose integration into a common European energy policy the authors of the agreement originally had in mind. Other states that have not ratified the Charter include Belarus, Iceland and oil- and gas-exporter Norway. The Transit Protocol – that part of the Energy Charter that Russia's criticism has been aimed at – provides for equal access by companies and third-party countries to Russian pipelines. This contradicts the decision of the Russian parliament to make Gazprom the only exporter of Russian gas. The EU, in contrast, wants to have the option of purchasing gas directly from Central Asian exporters, for example, Turkmenistan, and to transport it via the Russian network.

According to the Energy Charter, interrupting the transport of energy where a dispute before the relevant international mediation process has not yet been adjudicated, will in future be outlawed. Interrupting gas supplies, as was done by Gazprom in the dispute with Ukraine over the gas price, would no longer be possible.

Russia, in turn, believes that it has so far complied with the basic principles of the Energy Charter, such as the mutual opening of the markets and the protection of foreign investments – although the latter assumption could rightfully be challenged after the Russian government recently pressed foreign energy companies like BP and Shell to sell their shares in Russian energy projects to domestic competitors. This opinion, however, is also included in the Final Declaration of the G8 summit of summer 2006 and the joint declaration of the EU-Russian Summit that took place in autumn of the same year. Finally, the Russian government still wishes to conclude a new partnership and cooperation agreement with the EU in 2007. Chancellor Merkel announced in this regard that Russia had at least to accept the contents of the Energy Charter. The Polish government threatened that it would only put the new cooperation agreement into force once the Energy Charter was ratified by Russia.

Méthode Monnet or Putin method?

In October 2000, a new initiative to establish a permanent forum for discussing issues of mutual interest was undertaken. During an EU-Russia summit attended by heads of state and government, an energy dialogue was initiated and since maintained. During its first stocktaking exercise in December 2004, the EU Commission listed as objectives increasing Europe's energy security, diversifying and stabilizing the market and assisting with the sustainable development of Russia's energy resources. To summarize it in Brussels' words: The energy dialogue is to create a relationship of 'constructive interdependence'. The German federal government uses the term 'rapprochement by intertwinement', by which it wishes to continue the policy tradition towards the Eastern bloc during the Cold War. As it was expressed in a confidential strategy paper of the Ministry of Foreign Affairs on the German EU presidency in 2007, Russia must understand 'energy security as a cooperative coexistence of producer, transit and consumer states'. As long as Russia has not ratified the Energy Charter, energy-political cooperation should instead be regulated by a new basic agreement between the EU and Russia. The final objective would be a European-Russian free trade zone.

The EU has already agreed to Russia becoming a member of the WTO. In return, Russia ratified the Kyoto Protocol on climate protection. The decision to link the EU's approval to Russia's WTO membership and the ratification of the Kyoto Protocol by the Russian duma is a classic example of how the political negotiating process with the Russian government works. Topics that are only remotely related are politically connected. If a political agreement is achieved with the Kremlin, it might result in decisions by seemingly independent bodies, the parliament in this case.

The EU, which has so far favoured the Monnet economic and political approach, is confronted by the completely different 'Putin method' when dealing with Russia. Apparently, the Kremlin believes that the dispute over the control of the energy sector is a zero-sum game that you can only win if your opponent loses. Europe, on the other hand, believes in creating more common growth through economic integration, and more common security through political integration. Particularly in the environmental area, problems can only be solved by transnational cooperation; while the climate problem can only be tackled through worldwide cooperation.

Europe will only be able to withstand the political power play of Russia – which use its market power to push through its foreign policy targets without concern for anything – if it act as a united entity. Only then will important European interests, such as the protection of climate and the environment, or the concerns of the small Eastern European member states, not be ignored.

The European Neighbourhood Policy

The security and diversification of its energy supplies is the focus of the EU's strategic interest. The 'strategic energy partnership' with its neighbouring countries is therefore one of the pillars of EU foreign policy.

Domestically and through its gradual extension into the wider Europe, the EU has created a unique region of security and stability based on a positive development in a common market. European foreign policy aims to extend this region and export stability to its neighbouring countries as well. But both to the east and in the southern Mediterranean, the EU is surrounded by states going through economic as well as political crises. As part of its so-called Neighbourhood Policy, the EU aims at finalizing close cooperation agreements with all the states in Eastern Europe and the Mediterranean. What is to be achieved in the end is a pan-European free trade zone and close cooperation in political questions.

The EU imports energy from Russia, the Caspian region, North Africa and the Near East. Important energy transport routes go through other neighbouring countries, for example, Ukraine, Belarus and Georgia. However, in order to further develop economic relations with its neighbours, it is important that investments in the energy sector fit into a general economic strategy that allows the economies of the neighbouring states to identify strongly with the EU domestic market. Regional security interests play a role in this regard, as do democratic development and participation of the population. From an environmental perspective, it is about climate protection, clean air and water, and nature conservation.

Another Neighbourhood Policy objective is to promote the signing and ratification of regional and international environmental agreements and to ensure that these agreements are also implemented. After Russia ratified the Kyoto Protocol and the latter has now been put into force, implementation of the climate protection agreement has now become the centre of attention. Apart from the financial instruments, the EU emission trading system is the most important lever for pushing neighbouring countries towards a more sustainable use of their energy resources. For Eastern European countries, emission trading offers enormous incentives for modernizing their energy infrastructure. The promotion of energy efficiency and renewable energy is also to be part of the action plans. What is even more important, however, is that the strategic energy partnership with the neighbouring states does not only focus on the import of the two fossil fuels oil and gas, but also on the climate-political objectives of the Union. Agreements have, for instance, been signed with Russia and Ukraine to improve the safety of the local nuclear power stations.

Europe's energy Ostpolitik – the challenge of Ukraine

Ukraine is the most important transit country for energy imports from Russia to the EU. Without the involvement of Ukraine, a common European energy policy would therefore not be complete. At the same time, Ukraine itself depends on gas imports from Russia and other CIS states, the most important being Turkmenistan. All in all, there must be a decision on whether Ukraine is to be integrated into the domestic energy market of the EU over the long term, or whether it will remain in the dominion of Gazprom and the Kremlin.

Ukraine's current energy policy is a good example of how big the challenges are for modernizing energy policy in typical transformation and emerging markets. The energy-political challenges that Ukraine faces are enormous. Energy consumption per unit of economic output was 2.6 times higher than the worldwide average in 2005. This is due to the immense waste of energy by the heavy industry inherited from the Soviet Union as well as by private households. During the Soviet era, private consumers could get energy almost free of charge. This has changed, since now Russia's Gazprom determines the price. Many Russians, Ukrainians and Belarusians are therefore facing energy poverty, a completely new phenomenon. Either they pay the exorbitant costs of the gas bills during winter and have to save in other areas – or they stay cold.

Ukraine needs help from outside to invest in its dilapidated energy infrastructure and to reduce dependence on imports from Russia. After the existing pipeline network had been constructed, primarily with Russian capital, and Gazprom signed long-term supply and transit contracts with the Kuchma government in late 2004 – shortly before that government was ousted during the Orange Revolution – the EU offered further extension and modernization of the pipeline network through its financing programme TACIS. Now the EU and Ukraine are discussing a new partnership and cooperation agreement that provides for Ukraine's gradual integration into the domestic energy market of the EU. The EU, in turn, is deeply interested in making the transit country Ukraine part of the trans-European energy network and the common domestic market. With Ukraine on its side, the EU could act more confidently towards Russia.

Even through Ukraine is a country rich in resources, it has to import most of its energy. Even the hard coal from the Donez basin, the largest coal-mining area of the former Soviet Union, no longer suffices to fulfil domestic demand. Almost 80 per cent of the natural gas is imported from Russia and – via Russian pipelines – from Central Asia. The most important national energy resource of Ukraine is hard coal. However, two thirds of Ukrainian coal mines are not profitable. As the coal exploitation region of Donez had been part of the power base of the former Kuchma/Yanukovich government and the coal-mining industry is one of the major employers in the east of the country, the industry, which is controlled by a few oligarchs, receives major subsidies from the state. It will not be possible to discontinue these subsidies overnight, even though coal mining is accompanied by considerable safety and health risks as well as ecological threats.

More than 50 per cent of the steel plants in Ukraine, which are currently operating at full capacity since export is booming, still use the Siemens Martin process, which was developed in the middle of the 19th century. Ukraine has therefore one of the world's highest CO_2 emissions. For Western European power plant constructors, a potentially gigantic market exists here.

As avoiding CO_2 emissions in Ukraine – in particular in the coal and steel industry – is relatively easy and cheap, it would be profitable for the country to join the European emission trading system. In fact, Ukraine ratified the Kyoto Protocol even before Russia did. Western European companies could fulfil their obligations in terms of climate policies through comparably low investment in Ukraine. However, in order to introduce the complicated EU emission trading system, both the plant-specific energy consumption and the institutional prerequisites in Ukraine's environmental administration for monitoring such a trading system would have to be established.

The potential for renewable energy resources would also be enormous, but has hardly been used to date. Apart from wind and water power, Ukraine's major potential is the use of biomass. Traditionally, Ukraine was once the breadbasket of the Soviet Union. Today, large tracts of agricultural land are not cultivated, and agricultural products are not competitive on the global market. The restructuring of the energy and the agricultural sectors could thus go hand in hand, if Ukraine were to use biomass. Biogas could replace a part of the natural gas from Russia, and biomass power plants could replace electricity generated from nuclear power. Staff divisions of major German energy and automotive companies are already contemplating signing contracts with Ukraine in respect of agricultural land for cultivating crops for the production of ethanol and biodiesel.

The second domestic energy resource of Ukraine is uranium. According to the World Energy Council, domestic production in 1999 ranked tenth in the world with 3 per cent of worldwide production. In spite of this, additional uranium needs to be imported from Russia due to the importance of nuclear energy for national electricity generation. Ukraine would consequently not reduce its energy-political dependence on Russia by extending the national nuclear industry.

Nevertheless, the Ukrainian government plans to erect another 20 nuclear power stations by 2030. Last year, nuclear power stations contributed 46 per cent of electricity and 14 per cent of primary energy production of Ukraine. At the moment, 15 reactors are operating in the country. Khmelnitsky 2 and Rovno 4 (jointly also referred to as K2R4), both commissioned in 2004 to replace the Chernobyl plant that was shut down in 2000, are the youngest nuclear reactors in Europe. The construction of the K2R4 reactors was initially cofinanced by the European Bank for Reconstruction and Development (EBRD), with the political objective of shutting down the Chernobyl plant. In this regard, an internal study of the Bank came to the conclusion that the alternative construction of new gas power plants would have been considerably more cost-effective. When the EBRD eventually also applied stricter safety requirements, the Ukrainian government cancelled the agreement and completed both reactors with Russian and national financial support. Once the project was completed, using Russian nuclear technology, the EBRD and the European Commission had to finance the refitting of safety technology to the amount of €120 million.

The lesson of Chernobyl

Ukraine stands as a symbol for the abuse of nuclear energy, being the scene of the greatest catastrophe in the civil use of nuclear energy to date. When on 26 April 1986, reactor number four of the Chernobyl power plant got out of control, eight tons of radioactive material was released into the environment. The nuclear catastrophe of Chernobyl and the attempt of the Gorbachev government to cover up its real extent contributed to the downfall of the Soviet Union. According to UN data, 150,000km^2 in Ukraine, Belarus and Russia are still so radioactive that cultivating healthy crops is not possible. UN Secretary General Kofi Annan said in July 2004: 'At least three million children in Belarus, Ukraine and the Russian Federation require medical treatment.' Six million people still live in the affected areas. The agricultural damage due to health impairment and the loss of agriculture as source of income represent a significant long-term burden for all three countries. The sad commemoration day, which had its 20th anniversary in April 2006, should therefore be an occasion for the EU to talk to its Eastern European neighbours about a reorientation in their energy policy, which still focuses on nuclear energy. Chernobyl and the consequences also created an awareness of how small Europe and how big mutual dependence actually is – both in a positive and a negative sense. Energy-political decisions in one country have far-reaching consequences for immediate neighbours, and sometimes even for more distant countries. The lesson learned from this fiasco is that the EU can only influence important individual energy-political decisions of its neighbouring countries if it can offer these countries comprehensive economic integration. Ukraine's planned integration into the EU's domestic energy market would ensure that the Ukrainian energy industry would be orientated towards Europe over the long term and would adopt the safety standards customary in Europe.

The silk route of the 21st century

For years, Turkey has tried to strengthen its geostrategic position as energy export hub. The centrepiece of this strategy is the Baku-Tbilisi-Ceyhan (BTC) pipeline that was commissioned in July 2006. The BTC pipeline transports oil from the Caspian Sea via Azerbaijan and Georgia to the Turkish port of Ceyhan on the Mediterranean. Russia, Iran and Armenia are bypassed. In future, the pipelines from the oilfields of Kirkuk and Mosul in northern Iraq will also go to Ceyhan. Kazakhstan also wishes to export oil via the BTC pipeline. Turkish diplomats have already poetically referred to it as the 'silk route of the 21st century'.

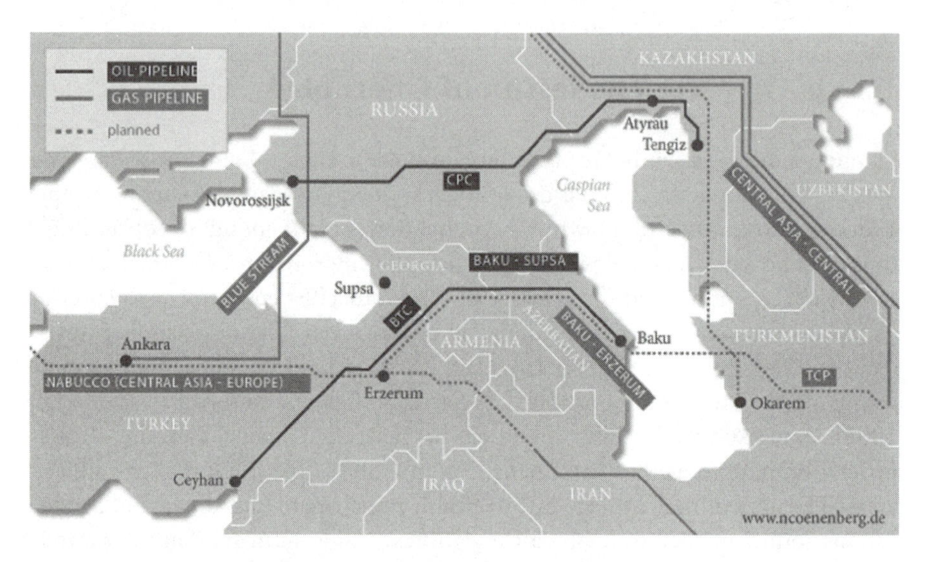

Figure 5.1 Oil and gas pipelines in the Caucasus region

Several gas pipelines, converging on Turkey's eastern border, are planned for the near future. An example is the Nabucco pipeline that will transport natural gas from Iran and Central Asia to the West. The construction of a natural gas pipeline with branches to the Georgian and Iranian borders and further to Austria, is also planned. The Western European market is to be supplied from there. The estimated cost of the pipeline is €4.6 billion. Planned to be completed by 2011, it will connect the EU with the Caspian and Iranian oilfields and thus reduce dependence on Russian imports. The project is one of the Trans-European Energy Networks supported by the European Commission and is secured by loans from the European Investment Bank (EIB), a national bank of the EU. Kazakhstan and Turkmenistan, who are currently directing their exports mainly via the pipeline network of monopolist Gazprom, could also tap into the supply lines of the Nabucco system. Egypt and other gas suppliers in the Near East have likewise indicated interest.

The key regional position of Turkey, and its plan to become the energy export hub to Europe, are important reasons why foreign policy strategists in European capitals are pushing for Turkey to join the EU. However, a number of security time bombs must first be defused to ensure that Turkey will be a stable partner for more energy security on the European continent. Apart from old border disputes with Greece in the Aegean, the Cypriot conflict and an as yet unresolved past of genocide against the Armenian minority, there is the civil war in the Kurdish south-east of the country. In the border region of Kurdistan, not far from the oilfields of Kirkuk and Mosul, Turkish, Iranian and American interests are in conflict. If Turkey wishes to join the EU it must solve the Kurdish problem peacefully, allowing political and cultural autonomy for the Kurds. It must also arrange a fair share of profits from the lucrative oil transit through the region, instead of securing it militarily, as if it were going through enemy territory.

While Turkey's future as energy hub might look bright, the situation at home is precarious as far as the country's own energy supply is concerned. Eighty per cent of Turkish natural gas imports come from Russia and Iran. The Ukrainian-Russian energy crisis early in 2006 therefore caused nervousness in Ankara as well. As part of its EU membership proceedings, Turkey should therefore be interested in becoming part of a powerful European negotiation bloc to help its dealings with the large gas exporters to its north and east.

The Southern Caucasus – Europe's forgotten neighbour

Since ancient times, it has been known that oil can be found on the peninsula of Apsheron, close to what is today Azerbaijan's capital Baku. In ancient Greece, the oil that emerged naturally on the surface was not only used as remedy and preservative, but was also the basis for the fire cult of Zarathustra. In the late 19th century, the Rockefeller and Rothschild families rediscovered the oil wealth of the Caspian Sea. If one travels from Baku southwards in the direction of Shirvan National Park and the Iranian border, one still passes the old wooden oil rigs submerged in mud, before reaching the construction sites of Western oil companies.

The southern Caucasus has once again become interesting for international politicians as energy producer, but also as an energy transport hub from Central Asia. A James Bond movie begins in the region – the following dialogue takes place:

M: 007, what do you know about the Caspian Sea?

James Bond: Caviar. Gigantic, wonderful Belugas. Firm, but tender.

(*M looks at him with indignation.*)

James Bond: The world's largest inland sea. A lot of oil. Hitler wanted the oil, but Stalin beat him.

M: It is about oil… Three pipelines go to the Black Sea and are threatened by unstable governments or a blockade by Russia.

James Bond: And the new pipeline project ... goes through Turkey directly to the Mediterranean Sea?

(*The World Is Not Enough*, Metro Goldwyn Mayer, 1999)

Agent 007 was far ahead of his contemporaries in the EU capital of Brussels. Even though Georgia, Armenia and Azerbaijan participate in European Neighbourhood Policy, a system of cooperation agreements between the EU and its neighbouring states, Brussels had almost forgotten about these three countries after the Soviet Union collapsed, this in spite of the fact that the Southern Caucasus has been in the centre of a geopolitical power struggle between the large neighbour Russia, the two regional powers Iran and Turkey, and the US. Immediately after the end of the Cold War, the US government recognized the three former Soviet republics' potential as an energy transit region. Azerbaijan is an important petroleum producer itself. In addition, natural gas was also found some years ago offshore of the Azerbaijani coast in the Caspian Sea.

The construction of the BTC pipeline was politically initiated and secured by the US, even though the majority of the investment funds came from European companies, for example, British Petroleum (BP). The authoritarian government of Azerbaijan is almost exclusively dependent on oil exports. Georgia benefits from the transit fees, while Armenia, which lives in a permanent state of war with Azerbaijan over the mountain enclave of Nagorno Karabach, was cut off from the development. The BTC pipeline project consequently not only stabilizes the authoritarian regime in Baku, but also intensifies the conflict of interest between the hostile neighbours Armenia and Azerbaijan.

Since the BTC pipeline was constructed, Russia has lost its dominating influence over the oil industry of the region. But the Russians play politics with their semi-monopoly on the gas supply to Georgia and Armenia. When supply pipelines were sabotaged in winter 2005/06, heating and industrial plants stopped operating in both countries.

If the EU really has a long-term interest in integrating the states of the Southern Caucasus into the common European market and, in the long run, into the EU itself, a political approach that values the region exclusively for its relevance as energy transit, will not suffice. In order to bring the states in the Southern Caucasus closer to Europe, the former must subject themselves to an extensive political and economic transformation process. This also includes the gradual resolution of numerous 'frozen' minority and territorial conflicts, which have made transnational regional cooperation almost impossible to date. The objective of this strategy must be the overall economic development of the region, which is only possible by overcoming dependence on oil. The region's wealth in resources must be used systematically to promote regional cooperation through the creation of a common infrastructure. This must also include Armenia and the approach of Iran to trans-Caucasian and trans-European energy networks.

Trans-European energy networks

Even if, over the long term, the EU must succeed in reducing its dependence on energy imports through more efficient energy utilization and renewable energy, it will still depend on imports for the foreseeable future. European politics must therefore aim at diversifying these imports as much as possible to prevent dependence on individual exporters. The trans-European energy networks are the centrepieces of this strategy.

Since the early 1990s, the EU Commission has pursued the goal of connecting the common European domestic market by so-called Trans-European Networks (TEN). While work is under way at full force to install data lines and build railway and highway networks throughout Europe, the trans-European energy networks are currently rather patchy. Since the electricity market was liberalized, only the high-voltage power line network has been developed so far.

But positive examples of a more penetrating energy-cooperation between the EU and its neighbours do exist. The EU has already signed an agreement on energy-political cooperation with the countries of former Yugoslavia. Over the long term, all countries of the Western Balkans are to become members of the EU. Before that can be achieved, however, the EU, as a first step, wishes to extend its domestic energy market towards south-east Europe. Modern power lines as well as new oil and gas pipelines for the transit from Turkey and the Black Sea to Western Europe are planned. EU companies should have the option of investing in a modern energy infrastructure in Yugoslavia. As a prerequisite, the EU and other international players, such as the World Bank, support the privatization of the energy supply companies of the countries in the Western Balkans. Eventually, regional cooperation in the politically divided Balkans is to be strengthened. The common use of renewable energy resources, for example, wind power at the coast and water power in the mountains, would be suitable options in this regard.

The objective of the TEN is not only to facilitate free electricity trade and compensate for seasonal bottlenecks in supply, but also to reduce dependence on imports. For instance, once the Baltic States have shut down the ramshackle nuclear reactor Ignalina in Lithuania, they will be vitally dependent on electricity imports from Russia and are therefore jointly planning the construction of a nuclear power station. Alternatively, the three states could be connected with the European electricity network by overland lines from Poland.

However, the crucial question is whether it will be possible to include Russia in the planning and the construction of the trans-European networks. Gas is still exported from Russia to the EU via the Russian state monopolist Gazprom. The construction of the export pipelines that connect Russia to the West has always been a political issue for both sides. When Germany, in the late 1970s, intended to import major quantities of natural gas from the Soviet Union for the first time, it led to conflict with the US who feared that its allies in Western Europe might become economically dependent on the Soviets. However, the German government was not only looking at economic interests, but also at a policy of détente: common economic interests should generate mutual political dependence, thus making any military dispute impossible over the long term. Eventually the so-called 'Friendship Pipeline' from Russia to the West was built. Since then other pipelines have been built that connect Russia and the West. However, the EU's objective is that no country may monopolize the energy infrastructure in the common European domain and thereby exert political and economic pressure

The Southern Caucasus, south-east Europe and Turkey have gained in strategic importance to the EU in the past few years because of the energy transit routes through these countries.

The Kyoto Protocol and European emission trading

In the Kyoto Treaty on climate protection, the EU undertook to reduce its greenhouse gas emissions by 8 per cent by 2012, using 1990 as base. The base year 1990 is of particular relevance to the EU because it also marks the collapse of the Soviet economic system in Central and Eastern Europe. In the years after that, energy consumption in the states of the former Eastern bloc decreased drastically. Eventually, many industrial plants were closed down with devastating consequences for the economy and the labour market in these countries, but with positive effects for the environment and climate protection. Today, the economy in countries such as Poland, the Czech Republic and Slovakia is recovering. The question is how increasing energy demand can once again be met.

To achieve its CO_2 reduction target, the EU has introduced a system of emission certificate trading. Each EU member state receives an allocated number of emission rights that it can transfer to its national industry. In a future phase of the trading system, emission permits will, however, be allocated centrally from Brussels. The operators of large industrial plants and power plants must participate. All participants can trade their allocated emission rights with each other. The market players in this regard are industrial companies; the state merely monitors the procedure and allocates the emission quota. The basic idea behind this is that investment in modern and energy-saving plants will be done where these are economically most effective. After all, the atmosphere does not care about where in the EU – or in the world – one ton of CO_2 is saved.

Emission trading has two effects. A Europe-wide threshold for the emission of CO_2 is defined. In real terms, the consumption of fossil fuels is thus rationed all over Europe. Previously limitless access to energy is now subject to control by the state. Because the EU has set a joint maximum threshold, its member states and its economic players are interconnected in a common emission management system. Emission trading thus contributes to the deepening of European integration in a very fundamental, material way. Collectivizing the responsibility for reducing CO_2 emissions could be compared to European coal production under the ECSC contract. Whereas the economy is booming and energy consumption is growing once again in Central European EU member states, led by Poland and the Czech Republic, energy consumption and thus CO_2 emissions in former Soviet republics Ukraine, Belarus and Moldavia – and naturally also in the Russian Federation itself – are far lower than in 1990. As part of the pan-European emission trading system, these countries could make money by selling certificates to those EU member states that struggle to achieve their climate protection goals by their own measures. In fact, according to World Bank calculations, Ukraine could earn a lot of money by selling emission rights to the EU for another two decades. The European emission trading system is an attractive offer for the EU's eastern neighbours, not only from an economic point of view. It could also be a first step towards the legal and political integration of Russia, Ukraine and other neighbouring states into the common domestic energy market of the EU. Unlike the European Energy Charter, which Russia does not want to sign at this stage, Russia has signed and ratified the Kyoto Protocol. Based on this, the energy-political cooperation between the EU and Russia could be developed further within a common legal framework.

Eurasia's future

Any forward-looking EU energy policy must look beyond Europe's immediate local concerns. The authors of the European Energy Charter therefore not only had Eastern Europe and Russia in mind, but also the whole of the Eurasian double continent. This is the great potential of this widely underestimated international treaty. Japan, for instance, has already ratified the Charter; Mongolia is also a member. However, they have made their ratification conditional on their large neighbour Russia also taking this step. China, South Korea and the Southern Asian states enjoy observer status. The alternative to converting the European Charter into a Eurasian Energy Charter would be a separate regional alliance by the growth economies of East Asia. This could follow the example of the European Coal and Steel Community, which was also the beginning of

European integration after World War II regarding energy-political issues. For Europe, it is important that any new regional alliance in East Asia does not develop as EU competition, but rather views itself as a partner.

Once again, the key to Eurasian cooperation is Russia. When the European Energy Charter was established, it was inspired by the idea of integrating Russia into the common European region. But not only has the Russian duma not ratified the agreement to date, Russian politicians now even doubt whether it was a good idea to sign the agreement back then in the 1990s. After all, the balance of power between Russia and the West has shifted since then because of rising oil and gas prices and the consolidation of the energy industry under the control of the Kremlin. Besides, Russia is busy gaining back its political power over the former Soviet republics in Central Asia using the lever of economic dominance and control of the transnational energy infrastructure. Natural gas exporting Turkmenistan, for example, is wholly dependent on the pipeline network of Russian monopolist Gazprom. We can assume that strategists in the Kremlin plan the same for the countries in Eastern Europe and in the Southern Caucasus over the long term.

This is the reason why Russia insists on negotiating a sub-agreement that would be more favourable for it before ratifying the Energy Charter in the duma. Essentially, it is all about who will control the transport of oil and gas from Central Asia and the Caspian basin via Russian territory to Europe. Europe is adamant in its demand for free access for Western companies and investors. Russia, on the other hand, wants its state-owned companies to assume control over the trans-European energy networks under construction.

The European Energy Charter, a fixed set of legal regulations for trade, investment and market access with common objectives and minimum standards for environmentally friendly energy use throughout Eurasia, remains a vision for the EU. All political counter-concepts that are being discussed are eventually aimed at splitting up the continent into different power blocs. Energy-political cooperation would thus become more difficult.

A negative example in this regard is the proposal of a so-called 'Energy NATO', suggested by the national conservative Polish government under former Prime Minister Kaczynski in 2006. This would include a solidarity clause analogous to Article 5 of the NATO Treaty, where only member states of the EU and of NATO would be accepted. This proposal is clearly aimed at Russia and would not be helpful for ensuring energy security or other economic and political interests of the EU member states. Ukraine's integration into a common domestic energy market, as desired by Poland, could also be realized by a partial takeover of the European laws and regulations – the so called *acquis communautaire* – and without any new agreement. The same applies to the states of the Western Balkans and other countries immediately bordering the EU. Along the same lines, the establishment of a NATO East, comprising Russia and its satellite states, is being discussed in Moscow. Similar to the Warsaw Pact and COMECON, the former economic alliance of the Eastern bloc, this organization would be controlled by Russia. The Red Army would not hold it together, rather the Russian oil and gas transport monopoly, and the resulting dependence of Eastern Europe, Central Asia and the Southern Caucasus on the power of the Kremlin would. Kaczynski's Energy NATO and the Moscow bloc would confront each other in a 'cold energy war'. It would be a horror scenario for security and stability in Europe and for a long-term common energy policy.

Even more problematic for Europe would be a Russian-Asian energy alliance that might emerge from the SCO. Russia and China, the two founder states of the SCO, are both rivals and partners as far as energy policy is concerned. The booming Chinese economy needs reliable oil and gas imports from Russia. Russia, in turn, is interested in a new, financially strong major customer. Besides, competition is good for business. For this reason, Russia's former President Putin, when he visited Beijing in spring 2006, also announced the construction of a new gas pipeline from Western Siberian oilfields towards China. The sensitive issue of this project is the fact that, to date, Europe has exclusively been supplied from the production areas in Western Siberia. Unless production quantities are sufficient to supply East Asia in addition to the long-term delivery contracts signed with Europe, Europe and China would for the first time become direct competitors over access to resources.

Here it is again: the return of geopolitics. Political alliances will, in future, primarily serve to secure access to vital resources and markets for their own economies. Geostrategists in Moscow, Beijing and Washington also wish to secure the energy future of their countries by military influence and, if necessary, military means. Can the EU keep it up? It should not – at least not at any rate or any price. It would be advisable for the EU to give preference to the Monnet method in its energy-political cooperation with its neighbours, rather than a mere geopolitical power struggle. Energy cooperation with Russia, the other Eastern European countries and the states of the Mediterranean must be based on international agreements and within the framework of joint institutions. The strategy should be the establishment of a legal basis for existing economic competition. The existing cooperation agreements with individual neighbouring states and the Kyoto Protocol would serve as a good foundation in this regard. The long-term goal of the EU must be a basic agreement on Europe-wide energy policy. The European Energy Charter already comes close to this vision. From a European perspective, there should not be a race for the last resources: scarce resources can only be used without conflict on the narrow European continent based on cooperation. However, Europe's energy future must include becoming less dependent on resources that are in any case running out; the interests of foreign and environmental policy, energy and climate security go hand in hand.

Chapter 6

Defending the Last Paradise

In the search for the last oil and gas reserves on the planet, both the environment and democracy are under pressure. The exploitation of natural resources is threatening the once pristine environment in the Arctic, on the oceans, and in tropical and Nordic forests. Routes are forced through mountains, ice and forests to facilitate the transport of oil and gas. Poisonous wastes and accidents threaten the indigenous fauna and flora, thus depriving traditional industries of their basis for existence. Where oil rules, democracy suffers. Violation of human rights and the disturbance of traditional ways of life of the indigenous population lead to social conflicts and political instability. Environmental groups and other NGOs that protest against this policy therefore defend not only natural features, but also democracy and freedom. Freedom, justice and democracy, in turn, are the absolute prerequisites to make any economic development that goes beyond quick oil wealth possible in the first place.

The pearl of Siberia

One of the greatest victories achieved by the Russian environmental movement in the past few years was the diversion of the planned trans-Siberian oil pipeline from the banks of Lake Baikal to an ecologically less sensitive route. As early as the 1980s, Soviet scientists, defenders of the cultural heritage and environmentalists started fighting for the protection of Lake Baikal. In the beginning, their cautious protest concentrated on the protection of a few symbolic species, for instance, the rare Baikal seal. Later on, this developed into an international environmental protest campaign that cleverly succeeded in playing off against each other the interests of the various parties involved in the dispute over Siberia's oil wealth.

The Baikal seal, chosen by environmentalists as the symbol of this Siberian Sea, is the only type of seal to live exclusively in fresh water. It probably descends from the ringed seal that lives everywhere in the Arctic. Nobody knows how it reached the interior of Siberia. After it had become almost extinct because of hunting during the first half of the 20th century, 84,000 animals were counted again in 2000. Apart from the ongoing illegal hunting, the seal is threatened by poison that is discharged into Lake Baikal and by the destruction of fish populations, above all the Baikal *sculpin* that lives at the bottom of the lake. It makes sense therefore that the seal is also the heraldic animal of Siberian environmental organization 'Ecological Baikal Wave'.

Lake Baikal is often referred to as 'the Pearl of Siberia'. It is one of the world's largest lakes, 650km long and up to 80km wide. It is also the oldest (25 million years) and the deepest (1700m) lake in the world. The maximum capacity of the lake is 23,000km^3 of water. This is more than the capacity of the Baltic Sea and 460 times the volume of the contents of Lake Constance on the Swiss/German/Austrian border. The catchment basin of the lake, including its inlets, comprises 1,487,480km^2, four times the size of Germany.

The environment of Lake Baikal has a unique flora and fauna: approximately two-thirds of the total 2500 animal and plant species are endemic, which means that they can only be found at this location. Apart from the Baikal seal and the Baikal *sculpin*, the *omul*, a type of salmon, and the *golomyanka* or Baikal oilfish, the world's deepest-living freshwater fish, can be found here. One of the factors favouring this unique fauna is the low water temperature of the lake, an annual average of only approximately 7°C at the surface. UNESCO, the scientific organization of the UN, compares its biodiversity to that of the Galapagos Islands and, in 1996, included Lake Baikal on its list of 'endangered world natural heritage sites'.

It is endangered because state-owned Russian oil company Transneft planned the world's longest oil pipeline from Irkutsk to Vladivostok in the Pacific passing 800m to the north of Lake Baikal. For a long time, the protests of Russia's national environment authority were ignored. The East Siberian pipeline will supply the Asian market, above all China, Korea and Japan, with Russian petroleum. To date, oil has been exported to the East exclusively by railway, and consequently not in quantities that would be economically viable.

Environmentalists, including the 'Ecological Baikal Wave', fear that the freshwater reservoir of Lake Baikal and its unique fauna might be endangered by accidents, natural catastrophes and terrorist attacks. Baikal Wave was founded during the era of perestroika to preserve the former natural monument of Lake Baikal from the rapid industrialization of eastern Siberia. In the early 1990s, the main problems included rapid deforestation and with it erosion of the banks, as well as a large wood pulp factory that discharged its waste water into the lake. Apart from the ecological impact, this threatened the economic livelihood of the local indigenous minorities that live primarily from fishing. If the gigantic wood pulp factory was symbolic of the large polluting projects of the Soviet era, Transneft's pipeline represents the 'New Russia' that wishes to restore its influence on world politics through oil and gas exports.

According to UNESCO, two-thirds of the route of the planned pipeline would go through seismically active territory. Transneft's reputation was also reason for concern. Between 1993 and 2001, the region of Irkutsk saw a total of six major leakages in pipelines and systems, during which a total of 42,000 tons of crude oil was discharged into the environment. These were the reasons why a pipeline to the south of Lake Baikal that had already been planned in the mid-1990s was not realized. Further along its route, the pipeline was also to pass the Nature Reserve of Kedrovaja Pad, one of the last habitats of the snow leopard.

In spring 2006, President Putin eventually decided that the pipeline should be built further to the north, away from the ecologically sensitive banks of Lake Baikal. It was not only ecological concerns that made him do so. The international environmental campaign organized by Japanese NGOs for saving Lake Baikal had cleverly taken advantage of the opposing political interests that had developed over the construction of the pipeline. Eventually, environmental groups provided oil exporter Russia and oil importer Japan with the right ecological arguments to reach a geostrategic decision opposing the energy interests of the third party in the trio, the Chinese.

The southern variant of the pipeline, which was rejected in the end, would automatically have gone to China; the northern variant ends in a Russian harbour. From there, the Japanese and Koreans can be supplied by ship. While estimates for the construction of the northern route suggest that it will be significantly more expensive, it leaves Russia with the flexibility of playing off East Asian oil importers against each other. As, once again, the Kremlin wishes to use oil exports as an instrument of its foreign policy, Transneft was instructed to pay the higher price for the route that goes through the difficult territory in Siberia's interior. Ecological arguments and the fate of Lake Baikal played a significant role in the massive lobby campaign of the Japanese government for the pipeline variant it favoured. President Putin announced his decision at a press conference before the G8 Summit in St Petersburg, thus publicly demonstrating to the world that environment and democracy are also respected in Russia.

The oil island of Sakhalin

Oil pipeline projects of the monopolist Transneft are sacrosanct. Any other interests are recklessly subordinated to the extension of the Eurasian pipeline network and the strategic goal of oil and gas exports. Examples in this regard include the exploration of the Arctic, the offshore production along the coasts of the Pacific, the Arctic Ocean and the Baltic Sea, and above all oil production on the Pacific island of Sakhalin.

Russia's new Arabia lies in the east of Siberia, on the Pacific island of Sakhalin, situated to the north of Japan. There, the Russian government and international energy companies are currently exploring several major oil and gas fields for export to Japan, China, Korea, and for supplying the global market. The supply pipe network that still needs to be built is to transport the Black Gold to China, Japan and the two Koreas.

Financial support for the project comes from the European Bank for Reconstruction and Development (EBRD), among others. When Anton Chekhov visited Sakhalin in 1893, he had the feeling of having reached the end of the world. 'From here, you cannot go further', he wrote in a letter home. For the oil industry as well, Sakhalin marks a new border. Small quantities of oil were already being produced on the island in early Soviet times, mostly by the prisoners of the far eastern penal camps. Today, however, focus is on offshore production along the island's coasts. Here, the oil industry is facing challenges previously unknown, under Arctic conditions, in pack ice and on high seas.

If the international companies that invest here can handle the challenges of Sakhalin successfully, the exploitation of other assumed fields in the Barents, Beaufort or Bering Seas, or in Northern Alaska and Canada would theoretically also be possible.

The Russian government has divided Sakhalin and its offshore territories into six exploration zones. The consortium for the exploration of Sakhalin-1 is managed by the American company Exxon. Exxon Neftegas Limited, an affiliate of ExxonMobil, is the operator of the Sakhalin-1 Project. Other participants in the Sakhalin-1 Consortium are two Russian companies, Sakhalinmorneftegas-Shelf and RN-Astra, the Japanese company Sakhalin Oil and Gas Development Co. Ltd, and India's ONGC Videsh Ltd.

Sakhalin-2 was initially operated by Royal Dutch Shell jointly with Japanese partners. It is now managed and operated by Gazprom lead consortium Sakhalin Energy Investment Company Ltd (Sakhalin Energy).

Fields 3 to 6 are to be awarded at a later stage. These licences are among the most attractive prizes that the Kremlin can offer international investors over the coming few years. Political considerations and concessions on the part of the foreign partners in other issues are certain to play a significant role in the decision. The concessions for Sakhalin that were awarded in the early Yeltsin years initially represented a gap in the export monopoly of state-owned Gazprom.

However, Sakhalin is also the test case for how the fragile Arctic environment, its marine life and the interests of the indigenous peoples of these regions will be handled. The exploration area Sakhalin-2 is particularly problematic from an ecological perspective.

Until a few years ago, Sakhalin was largely unaffected by industrial development. Big game species that have become rare in other areas cavort in the Nordic forest landscape of the island in a comparatively mild Pacific climate. 65,000 rivers and streams flow through the island and serve as breeding grounds for the Pacific salmon. Wildlife at the coast is even more abundant with large seal colonies and many sea birds, such as the rare Stellar sea eagle. The last 120 Western Pacific grey whales live in the sea off Sakhalin; this species had been believed extinct and was rediscovered there in the mid-1970s. Russian marine biologists, who have since then monitored the population, believe that the group includes only 20 females of reproductive age. The underwater pipeline of the major project Sakhalin-2 would lead directly through the grazing territories of the grey whales, who would not only be threatened by the destruction of underwater fauna and possible oil leakages, but also by the immense construction noise that would interfere with their sensitive sonar system in the next few years.

Sakhalin's beauty would be endangered by the entire oil production chain, which is to extend over the entire island. Significant volumes of pollution is discharged from the oil platforms into the marine environment and thus the feeding grounds of grey whales, seals and numerous ocean fish species. Ships not only disturb the peace, but also the routes of migrating marine species. According to current planning, construction rubble and excavation residues from drilling work on the Sakhalin-2 project are to be dumped directly in the bay of Aniva off the island.

All these concerns, raised by international environmentalists for years, have lately found attentive ears with the Russian environmental authorities, who are particularly concerned about the protection of grey whales. For this reason, the Moscow authority threatened in summer 2006 to cancel Royal Dutch Shell's licence to continue extending Sakhalin-2. However, the background to this previously unknown rigour on the part of the environmental authority is something else. The Kremlin wanted the operator consortium to sell shares to the Russian Gazprom group and to handle the export via the state monopolists. To prevent creating an impression of official arbitrariness, the environmental argument of the licensing department came exactly at the right time. Now that Gazprom has reconquered the Sakhalin project for the Russian state, it remains to be seen how strictly environmental legislation will be upheld.

However, the oil and gas projects are also a test case for the international banks involved in its financing, among these the public EBRD. The law requires them to adhere to more stringent regulations and environmental requirements in their financing business than those in force in a country such as Russia. So far, the EBRD has argued that its involvement in the project ensures the minimum standards in terms of transparency, public involvement and environmental protection. Consequently, European environmental organizations have the opportunity and the duty to demand that the tax-funded development bank complies with these regulations. The options that NGOs have to influence investment decisions via international financiers are described below.

The Courland Spit

The Courland Spit is situated at the other end of the Russian Empire from Sakhalin. Off the Courland Spit and in close proximity to the Lithuanian and Polish Baltic Sea coast, the Russian company Lukoil began drilling for oil in

2004. Even though the yield is expected to be modest, it threatens to destroy the world heritage site of the Courland Spit if there were only one major accident during production or transport of oil. In such an event, the project would not only threaten the Courland Spit, but the entire Baltic Sea. Already, pollution of the Baltic Sea is above average. As its ecosystem is highly sensitive, environmentalists support a general ban on oil production in this area. To date, Russian oil is transported in single-wall tankers with a high accident risk through the narrow fairways of the Baltic Sea. Lukoil has, in the meantime, given up a project for constructing a large oil refinery in Kaliningrad (the former Königsberg) following protests by civil society.

UNESCO declared the Courland Spit a world heritage site in 2000. The almost 100km-long sandbank separates the Courland Lagoon from the Baltic Sea, creating a unique habitat for marine fauna and flora. It also includes humid forests, pastures, swamps and reed areas making the Spit an important resting ground for migratory birds. For the enclave of Kaliningrad, which is isolated from Russia, ecotourism that has developed there over the past few years offers one of the few opportunities for economic development. In addition, the Lithuanian part of the Spit is one of the most popular bathing resorts in the country. German environmental organization Urgewald therefore writes: 'Numerous risks for the region are associated with oil production, the worst of these being the loss of the Courland Spit Nature Reserve. Even if the oil platform were to operate without any accidents, losses in the fishing industry must be expected, threatening several hundreds of jobs.'

Lukoil is one of the remaining private oil companies in Russia. The project offshore of the Courland Spit could therefore only be financed with the support of Western financiers. German banks such as HypoVereinsbank and Westdeutsche Landesbank (WestLB) are part of the bank consortium that granted Lukoil a loan of US$765 million. Following the Johannesburg World Summit on Sustainable Development in 2002, both banks signed the so-called Equator Principles, an undertaking by more than 40 major international banks to observe high environmental and social benchmarks. NGOs such as Urgewald are now urging international financiers of oil and gas projects to apply these standards.

The political issue of the Baltic Sea pipeline

Since former German Chancellor Gerhard Schröder was appointed as chairperson of the board of its operator consortium, the Baltic Sea pipeline planned by Russian Gazprom has become a political issue. The North Stream Gas Pipeline ('Baltic Sea Pipeline'), planned jointly by Russia and Germany, is controversial most of all because it cuts out Poland and the Baltic states. The expected environmental impact of this gigantic construction project is also frightening. Even though the construction of the pipeline was scheduled to begin in 2008, no comprehensive environmental impact assessment has yet been carried out. Sixty million cubic metres of submarine soil off the Baltic, Polish and German coasts would have to be excavated and 1.3 million tons of steel pipes would have to be installed for the construction of the 2100km-long pipeline from the gas fields of Juschno Russkoje in western Siberia to the Baltic Sea harbour Lubmin in Germany. The underwater channel in which the pipes are to be laid will be 30m wide. As the pipeline will go through shallow waters near the coast, it will cut through and destroy the spawning grounds of numerous fish species – and thus the economic base of the Baltic Sea fishing industry and the food base of sea birds. Shortly before reaching its destination, the projected route of the Baltic Sea pipeline will go through the Bay of Greifswald, called 'Bodden', and thus through the Baltic Nature Reserve of the coastal landscape of Pomerania. This nature reserve is a breeding ground and migration area for numerous rare bird species, for instance the sea eagle, the sea swallow and the crane.

Another silent danger has been lying at the bottom of the Baltic Sea since the two world wars. At the end of World War I, poisonous Russian and German gas grenades were dumped there. Parts of the Baltic Sea were mined during World War II. Sea maps of the north German Baltic coast mark 14 known mined areas alone. No such data is available for the Baltic states coast. In addition, it is not known what would happen to the flora and fauna of the fragile ecosystem of the Baltic Sea if one of the construction machines were to hit a grenade on the 2100km-long route, and poisonous gas were to flow into the sea.

Major projects worldwide

Russia is not the only country to unscrupulously exploit the last natural resources. In the US the pressure to explore the last paradise is also growing. For decades, American oil companies and local politicians have pursued the plan of producing oil in the Arctic National Wildlife Refuge (ANWR). Apart from the direct impact caused by the construction of production plants, far-reaching environmental consequences can also be expected from the necessary infrastructure and the transport by land, sea and air. The exploitation of oil reserves in the American Arctic is part of the energy plan of the Bush administration aiming at making the country less dependent on oil imports from the Near East. At least in the US, there is a powerful environmental movement and a free press that questions the sense of dubious projects, such as oil production in the ANWR. Since the opposition Democratic Party won the elections for the US Congress in November 2006, the project has been put on hold for now. The planned new production in Alaska would secure oil supply for the US only for a few additional months. However, the main problem is not insufficient supply of new production resources, but rather too high consumption in this country of boundless squandering of energy.

China's major oil and gas fields are situated in its Muslim western provinces and in Tibet. The exploration of new production fields, the construction of infrastructure and pipelines are therefore part of the comprehensive strategy of the leadership in Beijing to integrate these minority regions economically, and to bind them inextricably to the Han Chinese majority by importing labour. Major projects, such as the Sebei Lanzhou pipeline through Tibet, are therefore moved ahead through fragile mountainous regions with exclusion of the population. Obtaining reliable information from China on the environmental damage caused by these major energy projects is particularly difficult.

Even though major Western companies such as Shell and international financiers such as the World Bank are now under pressure from NGOs and the critical public, China's entry into the African oil business has resulted in a further decline in ecological standards and compliance with international laws and local regulations. When the IMF demanded that the government of oil-rich Angola introduced domestic reforms and a more democratic political approach before it would be granted a loan and support for the Angolan currency, the Chinese jumped in. China granted the government in Luanda loans of the same magnitude, with the sole condition that Chinese companies would get first option in tendering for new oilfields and the export of Angolan oil. The mingling of Chinese oil interests and the complete disregard for human rights is most evident in Sudan. China is the biggest investor in Sudan's growing oil sector. Through this, the Chinese support the Islamic government in Khartoum and its policies of human rights violations, slave trading and environmental destruction in the non-Muslim south of the country. Europeans and the US therefore demand that China must comply with certain minimum standards in the areas of the environment and human rights when investing internationally.

Many petroleum-exporting countries in Africa, Latin America or Central Asia suffer from economic stagnation, dire poverty, destruction of the environment and violent conflicts. The profit from the lucrative energy business goes to national elites and oil companies; very little money is reinvested in the affected regions. What remains are a destroyed environment and permanently damaged social structures.

Black gold from the dark continent

One of the first major projects to symbolize the economic rise of Africa as a resource provider of the world in the early 1990s is the Chad–Cameroon pipeline. As a model project, it has failed. Nevertheless, valuable conclusions can be drawn from the genesis of this project, which the World Bank and the European Investment Bank once supported enthusiastically.

When oil was discovered in the Sahel state of Chad, a country with no access to the ocean or any other transregional traffic connections, the question was not only how technically to connect the country to the global energy market, but also who would pay for it. Chad and neighbouring Cameroon are among the world's poorest countries. A civil war has been going on in Chad for several decades, also incited by neighbouring countries Libya and Sudan. According to Transparency International, Cameroon is one of the most corrupt countries in the world, ranking 138th of 163 in Transparency International's widely accepted Corruption Perception Index from November 2006. Therefore, when the World Bank was approached to finance the exploration of the oilfields in southern Chad, the construction of a more than 1000km-long pipeline and an offshore terminal for shipping oil to the global market, its development experts were initially sceptical. According to its statutes, the World Bank's mission is to help developing countries and fight poverty. However, it seemed doubtful whether the proceeds from crude oil production would benefit the poor in these two autocratic countries marred by corruption. In cooperation with the three major companies Exxon, Chevron and Malaysia's Petronas (also involved in the project financing), the EBRD, and advised by environmental and development organizations, it was decided that the loan would be conditional. The plan was that 10 per cent of the oil money was to be paid into a future fund; another 75 per cent would go to development projects for education, health and regional development, and only 12 per cent to the general budget of the two countries. A number of international monitoring committees were supervising the whole thing.

However, the first problems were already encountered during the construction of the pipeline. Its route went through fertile agricultural areas in southern Chad, and even through previously virgin rainforest and the territory of the indigenous Pygmy group. Apart from displacing the local population living there, agricultural land was destroyed and the water quality of adjacent rivers deteriorated.

When first profits started flowing, the regime of Chad, involved in a civil war with radical Islamic rebels, redirected funds from the education and health fund and used it to buy weapons. This was based on a change in the law made on the President's instructions, according to whom security expenses were also part of the country's development. The World Bank then froze some of the accounts of Chad, but released the funds again following negotiations, to ensure that at least part of the money was spent on development measures as originally planned

In the meantime, Chad's negotiation position vis-à-vis the bank has improved significantly. Higher oil prices made Chad less dependent on the international community's goodwill. Since the civil war in neighbouring Darfur spilt over into Chad, the West – primarily former colonial power France and the US

– are interested in stabilizing the regime. If Western oil companies wish to cancel the deal, Malaysian Petronas would remain. China's state-owned oil companies, already active in neighbouring Sudan, have their weapons at the ready as well.

Nevertheless, the Chad–Cameroon model was the first to experience how well conceived distribution and monitoring mechanisms can ensure that profits from a major publicly funded investment plan can benefit not only a few rich, but also the many poor. International environmental and development groups had been critical of the whole project, believing that it came too early for both countries, and arguing that in any case it should be linked to political reforms. In spite of this, it succeeded in ensuring that at least part of the oil profits went to education and health and not to Swiss bank accounts.

The law of petropolitics

American columnist Thomas Friedman has claimed that oil and democracy cannot coexist. In spring 2006, he published his theory in the magazine *Foreign Policy*, under the title 'The first law of petropolitics'. According to this idea, an increasing oil price will be accompanied by a decline in democracy in countries such as Nigeria, Venezuela, Russia and Iran. The new wealth is distributed unequally and is used by a few profiteers in the economic and political elites for strengthening the state suppression apparatus. However, some countries have succeeded over the long term in investing the wealth from natural resources in social development or using it for long-term goals. International initiatives demand at least more transparency so that multinational companies cannot become part of the crimes that local dictators commit against democracy and human rights.

Since the 1970s, the phenomenon has been described as the 'resource curse'; sometimes also as the 'Dutch disease'. When, in the 1960s, natural gas was found in the Netherlands, this presumed blessing led to an economic imbalance from which the Netherlands only recovered decades later. The export boom that was solely based on the production of natural gas led to the Dutch Guilder being overvalued. However, as the production and export capability of other industries declined, imports increased at the same time. Consequently, the presumed blessing of gas wealth led to a situation where the modernization of the Dutch economy in other areas stagnated and the country paid the price of high inflation, unemployment and the emigration of entire industries. However, the well-established democracy of the Netherlands and its legal system did not suffer. Therefore, the Netherlands was only infected by a mild form of the 'resource curse'.

In developing countries where the exploration of oil, gas and other natural resources make up most of the state income, there is the danger of generating economies merely based on yield. In economics, licence fees or taxes that the governments earn from the exploration of natural resources are referred to as yield, which the state earns from the ownership of or political control over the resources without having to invest in them. This seemingly effortless income tempts political elites to enrich themselves unrestrictedly. Instead of focusing on the long-term development of economy and infrastructure, investment is exclusively made in the exploration of oil and other natural resources. Because the government has the central hold on income from natural resources, authoritarian and centralistic tendencies are strengthened. Eventually, income from oil is only used for two types of investment: a villa on Lake Geneva and the police apparatus for suppression at home. Three negative examples from Africa, Latin America and Asia prove this theory.

The resource curse – Nigeria, Venezuela, Turkmenistan

At the height of the oil boom in the mid-1970s and during the time of the first oil crisis and rising prices, Nigeria had already been touted as Africa's Brazil: an economically strong, populous country that would soon assume a key position on the African continent. This was followed by almost 30 years of military rule, ethnic conflicts and constant economic decline. Today, two-thirds of Nigerians live in poverty; power is regularly cut off in the capital Lagos; petrol is only sporadically available at the state-controlled filling stations, despite the fact that Nigeria is still the major African exporter of crude oil.

It was not merely a case of bad luck for Nigeria. The sudden oil wealth shortly after independence made politicians and the economy of the country focus exclusively on the division of oil income. For years, Nigeria has been high on Transparency International's list of the most corrupt countries in the world. Corruption and inefficiency rule in the state-owned oil production company Nigerian National Petroleum Corporation. The lucrative positions in the oil industry are not awarded according to skills, but rather political alliance, or are simply sold to the highest bidder. Therefore, the infrastructure of the state-controlled oil industry has been deteriorating noticeably. For years, the oil production of international major companies in the Niger delta has only taken place with great protection from national security forces. The local population fights expropriation and the destruction of agriculture and the fishing industry by massive pollution, and also the fact that the entire income from oil goes into the pockets of corrupt officials and the military. There is nothing left for regional economic development, for schools and hospitals.

The international community started becoming aware of the intolerable conditions in the Niger delta when author Ken Saro-Wiwa and eight other human rights activists from the ethnic group of the Ogoni were executed in 1995 by the former Nigerian military junta because of alleged terrorist attacks against oil transport pipelines. Saro-Wiwa had also protested against the ecologically destructive production practices of Royal Dutch Shell. Shell, in turn, refused to use its substantial influence to obtain a pardon for the convicted human rights activists. Even Nigeria's conversion from a military dictatorship to a democracy did not end ethnic and social conflicts in the oil production regions. Local village communities illegally tap pipelines, and this causes frequent accidents and explosions with high fatalities. Because people are bitter about the unjust distribution of the oil wealth, there are frequent acts of sabotage causing further danger to people and the environment. Since several employees of international oil companies were abducted and only released after high ransoms had been paid, the oil multinationals have moved their oilrigs offshore. Even though production costs there are naturally higher, security expenses are less. Moreover, Shell and other oil multinationals ceased using Nigerian police for guarding their plants long ago, and instead employ private mercenary troops. These include international security firms, such as Erinys International, who offer a whole range of 'special services' from personal security to fighting against terrorism. Smaller companies that cannot afford expensive sentries have, in the meantime, withdrawn from the Nigerian oil business.

Nigeria is sitting on a powder keg because its radical Islamic movement has grown during the time of military dictatorship. Al Qaeda and other terrorist networks from Arabia are also recruiting in Nigeria. Their ideas of a holy war against the West and its democracy are becoming increasingly popular among the masses of young unemployed and socially discontent.

In 1999, former air force colonel Hugo Chavez was elected as Venezuela's president for the first time. Since then, democracy in Venezuela has been declining, according to the assessment of American NGO Freedom House, which has developed a worldwide index for the observation of human rights. Freedom House contends that, under Chavez, freedom of the press and the rights of opposition parties, NGOs and trade unions have been restricted massively. Transparency International currently ranks Venezuela at 130 on its global corruption index, sharing this place with Kyrgyzstan and the Congo. Nevertheless, the Venezuelan president is popular in his country. As long as oil income increases and is used to finance social programmes of the state, the political situation remains stable.

It happened once before, after the worldwide oil crisis in 1973, that the country's income from petroleum export skyrocketed. The Venezuelan government used the income for generous social grants, but neglected economic development. When the oil price fell significantly again after 1983, this income declined. The fact that no investment had been made in other industries that could have compensated for the drastically decreasing earnings from petroleum, combined with growing foreign debt, gave rise to persistent economic crises.

Today, oil prices are rising once again, and Chavez is busy making the same mistake. Besides, he combines a paternalistic domestic social policy with an aggressive foreign policy aimed against the US, by which he tries to reshuffle the political situation in Latin America.

When, in 2002, the opposition attempted a coup to oust Chavez, support from the US was obvious. Since then, Chavez has warned the US against interfering in the domestic affairs of the country, threatening to stop oil supplies. After all, 15 per cent of US oil imports come from Venezuela. Chavez also uses his oil wealth to support leftist, anti-American political movements in other nations in Latin America. Thanks to his support, the new Bolivian President Evo Morales came to power. One of Morales' first official acts was to nationalize the Bolivian gas industry and send out the military to occupy its plants. Bolivia's neighbour Brazil is particularly upset about this as the state-controlled Brazilian energy company Petrobras has invested heavily in Bolivia. Since then, there have been two development models in Latin America: the social democratic governments of Brazil and Chile, striving for democratization at home and the integration of their economies into the global market on the one hand; and the Venezuelan model of resource nationalization and the formation of a regional block against the US on the other hand.

Of the former republics of the Soviet Union, only a few have succeeded in their transition to democracy in the Western sense. One of the most authoritarian countries in the world and a dictatorship of the most bizarre type is Turkmenistan in Central Asia. The late President Saparmurat Niyasov, a former Soviet functionary, appointed himself as the 'Turkmenbashi', the leader of all Turkmen. Not only did he govern despotically and absolutely, he also ordered a well-contrived cult around his persona. Monuments of the Turkmenbashi, some of them gold-plated, can be found everywhere. Even the calendar months were renamed after the President and his family members. President Niyasov died in September 2006 and, after an election in February 2007, was succeeded by Kurbanguly Berdymuhamedov. The new president has promised to continue the policies of his predecessor, but also to introduce reforms, including unlimited access to the internet, better education and higher pensions.

Together with North Korea, Burma and Sudan, Turkmenistan is one of the world's most repressive dictatorships. In spite of this, international protests are limited. This is because Turkmenistan sits on the third largest natural gas reserves in the world. Therefore, its neighbours fawn on the country. US human rights organization Freedom House reported that income from exports amounting to US$1.3 billion per year went directly into the presidential fund of dictator Niyasov. China and Russia supported the Turkmen dictatorship politically and supply the Turkmenbashi with weapons and police equipment for its apparatus of suppression. After the terror attacks of 11 September 2001, Turkmenistan, which borders on Afghanistan, also became an ally of the US in the fight against terrorism for a short while. Since then, the democratic opposition in the country is suppressed even more brutally under the banner of the fight against alleged Islamic terrorism.

At present, Turkmenistan's natural gas is exported to Europe via Russian pipelines. This cheap gas subsidizes the purchase prices in Belarus and Ukraine. By these subsidies, which might be withdrawn at any time, Russia keeps its political influence and can threaten to destabilize the governments in Kiev and Minsk. Western Europe and Germany also buy gas from Turkmenistan. Once the Nabucco pipeline, which is to transport natural gas from Central Asia via the hub Turkey to Western Europe, is completed in 2011, the share of Turkmenistan natural gas for the EU is to increase even more. Even though consumers in the West are unaware of this, the suppression of the Turkmenistan people is also paid for in euros.

A comedy of power

Like oil and water, oil and democracy do not mix well. This applies to not only the corrupt despotic systems in Central Asia or the failed states in Africa, but also to apparently well-functioning democracies of the Western world.

The corruption scandal surrounding state-owned French oil company Elf Aquitaine, in which the German Federal Government also played an infamous role, is a good example in this regard. After World War I, when access to oil and petrol for the navy, aircraft and goods vehicles turned out to be decisive for the outcome of the war, France decided to implement a national oil policy. The French secured a lucrative share of the oilfields in Mesopotamia that formed part of the insolvency of the fallen Ottoman Empire. Until the eve of World War II, 40 per cent of French oil consumption came from Iraq. Then reserves were discovered in the Aquitaine region in southern France. From a number of existing companies the French government eventually founded the oil company Elf Aquitaine in 1976. The logo of the Elf group, which could be seen above the filling stations of the company, consists of a stylized drill and the French flag. From the outset, Elf also functioned as an instrument of French industrial and foreign policies.

In the early 1960s, French media reported that Elf Aquitaine was paying bribes to African politicians. This was about access to national production and distribution rights, but also about other French political objectives in Africa. For the government in Paris, the national oil company offered the ideal camouflage and the financial resources for political operations in areas of French interest in Africa. The company operated as a type of national intelligence service regarding the economic and political situation in the former colonies. Funds channelled via Elf were used for financing pro-France politicians and for obtaining weapons for civil war parties and militia that were on France's side. With the help of Omar Bongo, the President of Gabon, France even gained access to the innermost functioning of petroleum cartel OPEC. Bribes paid to Bongo and his family paid off by obtaining literally priceless information on internal price negotiations of OPEC.

What Elf demonstrated between the 1960s and the 1980s reads like the script for today's shirtsleeve approach of the Chinese on the African oil markets. US and other Western oil companies have not behaved fundamentally differently. So the criticism by the West with regard to China's investment in the oil wealth of African dictatorships such as Sudan leaves a bad taste in the mouth. In Germany, Elf Aquitaine purchased the Leuna factory in the formerly East German chemical triangle after reunification. The circumstances have never been fully clarified. As a lucrative additional deal, Elf was also allowed to take over the East German filling station chain Minol. Bribes were allegedly paid to the Socialist party of French President Mitterrand and to the German Christian Democratic Union (CDU). The CDU possibly used the funds from Elf to finance its successful campaign in the 1990 German parliamentary elections. The role of former Chancellor Helmut Kohl in the affair has not been clarified. In any case, the purchase of the Leuna factories and the modernization of the ecologically polluting chemical industry cost the German taxpayer billions in subsidies. The background in this regard was that Kohl had promised during the election campaign that the centrepiece of the East German chemical triangle, the Leuna factory, would be continued under any circumstances.

Funds from the company's cash box were not always handled properly. Substantial amounts landed in the pockets of French politicians and intermediaries from the business sector. The Elf phenomenon is described and ridiculed in Claude Chabrol's film, *A Comedy of Power*. In it, Isabelle Huppert plays an investigating judge modelled on the real person Eva Joly, who uncovered the kickback scandals around Elf. Elf Aquitaine's initial public floatation was in 1991. In 2000, the company eventually merged with another major oil company in France to become TotalFinaElf. Even though the French state's influence on this new company is no longer absolute, the objective of the merger to a homogeneous French oil giant was – and is – to retain a powerful instrument of French industrial and foreign policies in international competition.

Investing for the future

Not all states that have become rich through oil waste their wealth and decline into dictatorships. Norway is the only net oil exporter in Western Europe. The country put its oil income in an investment fund and has also invested in alternative energy for the future, such as wind and hydropower. In the US state of

Alaska, payments are made into a future fund that is to secure state expenditure even after proceeds from oil end. Besides, all citizens receive basic benefits from the state. Norway and the US, as was the Netherlands before it was infected with the Dutch disease, had been democracies before oil was found on their territories. Even though oil production has led to massive environmental damage, from the accident of the tanker *Exxon Valdez* to the destruction of grazing grounds for the caribou herds, it was at least decided by parliament what the national income should be used for. This model, where future funds are to ensure that future generations will also benefit from the yield once earned from natural resources, has also been applied in the African nation of Botswana. Botswana does not have oil, but diamonds. In addition, Botswana is one of the oldest and most stable democracies on the African continent.

From competition to cooperation

Competition over resources may create conflicts. However, it can also be an incentive for solving these conflicts in a cooperative manner through joint resource management. The environment and security initiative of the Organization for Security and Cooperation in Europe (OSCE) and the environment programme of the United Nations (UNEP) collect successful cases of resource management that have contributed to security and stability, so that the experience gained can be transferred to other troubled regions. Two regions where Germany development cooperation is actively involved in the civil prevention of resource crises are the southern Caucasus and the Altai Mountains.

The Altai Mountains extend over the borders of Russia, China, Kazakhstan and Mongolia. Russia's President Putin, when he visited Beijing in spring 2006, announced the construction of a new gas pipeline from west Siberian gas fields towards China, passing the short Russian-Chinese border section. This decision might potentially result in conflicts in three ways. First, gas from the fields in western Siberia, which to date have exclusively supplied Europe, would, for the first time, be directed eastwards. Russia would thus carry out its threat to play the Europeans and Asians off against each other on the gas market. Second, the pipeline, as the largest economic project in the region, would bypass the two adjoining states Kazakhstan and Mongolia. Neither neighbour would benefit from the investment or jobs, nor from a connection to the new gas supply line. In this regard, the conflict is similar to the dispute surrounding the Baltic Sea Pipeline that literally cuts out the Baltic States and Poland on its route from Russia to Germany. Finally, the pipeline project would cut through an important nature reserve in the Altai Mountains.

This is exactly where a model project of the German Federal Nature Conservation Agency (Bundesamt für Naturschutz or BfN) comes in. In negotiations with the four neighbouring states, the BfN has tried for years to create a transnational biosphere reserve. The goal is to preserve the most important natural treasures of the Altai, and simultaneously to offer better economic prospects for this poor and underdeveloped region by sustainable agricultural use and ecotourism. This is based on the idea that transnational cooperation on issues of nature conservation will create a culture of cooperation that goes beyond these four neighbouring states. The objective is improved regional cooperation and thus political stability. Even though the Chinese–Russian pipeline project has initially caused nervousness and additional tensions, it could be planned in such a manner that all neighbouring states could benefit from it. Changing its route, possibly through one of the two neighbouring states, would ensure a more environmentally friendly major project. Branch supply routes could supply the region with the relatively environmentally friendly natural gas. The joint planning process would not only bring environmental and planning authorities, but also political decision makers on all sides of the border region closer together. Finally, all the countries involved would also have to cooperate in other political areas, for example, the protection of the new gas pipeline against terrorist attacks.

Another troubled region where interests of environmental, energy and security policies clash is the southern Caucasus. Even though the three former Soviet republics of Georgia, Armenia and Azerbaijan do not share the same language or culture, they are all struggling with the same problems. The region is situated between the rival powers Russia, Turkey and Iran. All three countries are experiencing minority problems and secessionist movements. The economic situation has deteriorated dramatically since the collapse of the Soviet Union, membership of the EU that was hoped for is therefore moving beyond reach. The most serious conflict in the region is the dispute between Armenia and Azerbaijan over the mountain region of Nagorno Karabach. Both countries have been confronting each other without reconciliation since Armenia occupied this region in a bloody campaign in the early 1990s. One of the few hopeful signs is the cooperation of eight mountain villages on both sides of the border under the auspices of nature conservation and sustainable development. This project was initiated by the German Ministry for the Environment according to the model of the Alpine Convention that regulates protection and utilization among a large number of Alpine countries. This transnational cooperation in a seemingly remote region is gaining in relevance because the southern Caucasus has once again become the centre of geopolitical attention as a transit region for energy.

The Baku-Tbilisi-Ceyhan (BTC) pipeline, which was ceremonially inaugurated on 13 July 2006, is the largest energy infrastructure project of the 21st century so far. Through it, Turkey aspires to become the energy hub between Europe, the Near East and Central Asia. Plans are under way for other projects that are to pump oil and gas from these regions – bypassing Russia – to Europe. The BTC pipeline will pump oil from the oil port of Baku in Azerbaijan via

Tbilisi in Georgia to the Turkish port of Ceyhan on the Mediterranean. What is noteworthy about this three-country project that, by the way, is mostly backed by US economic interests, is who is excluded. Not only does the pipeline bypass America's geostrategic opponents Russia and Iran, but also Armenia – that is in conflict with both Azerbaijan and Turkey. Even though the pipeline connects three countries economically and politically, it excludes another three.

Besides, the selected route is also problematic from an ecological and social perspective. The Caucasus and northern Turkey are among the world's seismically most active regions. Should there be a major earthquake, there is the risk that the pipeline could break resulting in an environmental catastrophe due to possible oil spillage. Nobody knows how long it would take to repair the pipeline in the virtually impassable territory of the Caucasus, or whether such an earthquake would not bury and destroy the last supply lines. Naturally, the same problem would emerge in the event of a terrorist attack.

The pipeline cuts through one of the most important national parks of the Caucasus, one of the most species-rich regions in the world. The Caucasus is one of the Vavilov Centres – named after the Russian biologist – that produces the most biodiversity in the world. The country's most important potable water aquifer is also situated on the territory of the Borjomi National Park in Georgia. Not only does the water from the Borjomi supply the capital Tbilisi, but it is also one of Georgia's important export products. In future, one would have to say 'oil instead of water'.

The southern Caucasus is one of the most unstable crisis regions in the world. Proceeds from the pipeline construction are distributed very unequally. While Baku is developing into a glittering oil metropolis, unemployment and poverty are still the order of the day in the mountainous regions of Azerbaijan.

The cooperation project of the German Ministry for the Environment in the border region between Armenia and Azerbaijan is thus one of the few signs of hope in a region where otherwise conflict is rife. This is exactly why transnational resource management and joint economic regional development must be the focus of future policies. Otherwise, the dangerous mixture of competition between great powers, ethnic conflicts, economic underdevelopment and petronationalism that is brewing in the southern Caucasus will soon lead to a political explosion like the one in the Balkans in the mid-1990s.

Environmental peacemakers

Not only state players, but also NGOs and indigenous peoples' organizations contribute to finding cooperative solutions to resource conflicts.

One of the most innovative interstate organizations is the Arctic Council. Its members include not only the states bordering the Arctic, but also representatives of the indigenous peoples who live in the far north and had used it in an environmentally friendly manner even before any state borders existed in this region. The new competition over the last oil and gas resources has also reached the Arctic. For Inuit, Sami, Laplanders and the indigenous inhabitants of Siberia, the new oil boom in the north means not only a colonization of their home country by international oil and gas companies, but also a threat to their means of existence because of air and water pollution, and damage to the fragile local flora and fauna. The inhabitants of the Arctic are also among the first who will experience global climate change directly, and not just from newspaper and television reports. Ice fishing for seals is no longer possible because the ice off the Arctic coasts is disappearing. At the turn of the millennium, the north polar area completely melted for the first time in summer. It will soon be possible to use the waterways north of Siberia and Canada throughout the year. Thus, nothing obstructs the transport of oil and liquefied gas to the global market. The Arctic Council and its advisory committee for indigenous people see themselves as critics and first witnesses of the climate catastrophe that, in the end, threatens to change the living conditions of all of humanity. Like the canary in a cage that in the past warned the miners of escaping mine gas, the Inuits are now the first to demonstrate what the real threat of climate change looks like beyond statistical calculations and colourful graphs.

The organizations of the various indigenous peoples and the lobby groups of ethnic minorities focus attention on social and ecological destruction due to the exploitation of natural resources, not only in the Arctic, but also on Sakhalin, in Lake Baikal, Tibet and the Niger delta. However, the major contribution to enlightenment and thus rethinking comes from globally operating NGOS. The organization Bankwatch, for example, scrutinizes the loan allocation of international major banks, such as the World Bank and the EBRD. Between 1992 and 2005, the World Bank spent US$11 billion on projects for the exploration of fossil energies, US$4 billion of these on the extension of oil production. Eighty per cent of these World Bank funds were used for projects that promote the export of fossil fuels from developing to industrialized countries, thus further increasing the mutual dependence on the narcotic oil. In addition, the EBRD used a major part of its loans for oil and gas projects. As World Bank and EBRD loans reduce the political risk for private banks, they often have a catalyst function that attracts additional funds from private money markets. NGOs such as Bankwatch have long been critical of the situation where a major part of the energy-political projects of international development banks is spent on coal, oil and gas instead of on renewable energy or energy-saving projects and the construction of a modern, environmentally friendly energy supply structure in developing countries. The *Extractive Industries Review*, an internal study commissioned by the World Bank in the late 1990s concerning its projects in the area of natural resource exploration, came to the conclusion that the promotion of fossil fuel projects would not lead to long-term development successes and was therefore no longer purposeful. The group of experts recommended that the bank should put a moratorium on the support for coal mining and should discontinue crude oil projects. However, the shareholders of the World Bank, who do good business with precisely these projects, have so far refused to implement the recommendations of the *Extractive Industries Review*. The World Bank's board comprises mostly its largest shareholders, thus Western industrialized countries like the US and Germany, and also Japan. In addition, the bank's clients, the governments of developing countries, often insist on resource exploration projects of the old style. One wonders whether the politicians and high officials that decide on such projects in some developing countries have the interests of their country or rather their own bank accounts in mind.

German NGO Urgewald is only one of many that have uncovered through diligent research which financial interests lie behind individual major projects and how the ecological and social consequences of natural resources exploitation are covered up. The aforementioned example of oil drilling by Lukoil off the Courland Spit is only one of the scandals that Urgewald has uncovered and brought to the media's attention. Internationally connected and professionally organized environmental organizations work together with local partners and support them in the fight against overwhelming economic and political interests. For example, the Japanese branch of worldwide environmental organization Friends of the Earth supports local environmental initiatives on Sakhalin, in Indonesia and the Philippines, and at the same time keeps a close eye on Japanese and other international oil companies.

One of the most important steps towards a democratic decision on resource politics is the creation of transparency. The Open Society Institute of former major speculator-turned-philanthropist George Soros founded the initiative 'Publish What You Pay'. It requires all oil companies worldwide to disclose their payments to the governments in the countries where they operate. The Extractive Industries Transparency Initiative (EITI), initiated by former British Prime Minister Tony Blair at the Johannesburg World Summit for Sustainable Development in 2002, aims at bringing governments, companies and civil societies together at one table when major projects are implemented. Win–win situations are aimed at, in that all sides benefit from the planned major projects. Today, the World Bank also scrutinizes its loan business in the oil, gas and mining sector as part of its *Extractive Industries Review*. A report by former Indonesian Minister of Environment Emil Salim to the bank even demanded calling a halt to supporting fossil fuel projects altogether by 2008 and using the majority of the funds for projects promoting energy efficiency and renewable energy. Transparency is thus an indispensable step on the route to democratization of our energy policy, even if it cannot replace a basic rethinking towards environmentally friendly and renewable energy resources.

Development as freedom

One of the masterminds of the globalization debate and passionate defender of a democratic economic order is the economist Amartya Sen, of Indian origin, who teaches at Harvard University. In 1998, he was awarded the Nobel Prize for Economics for his work on welfare economy and the theory of economic development. In his book, *Development as Freedom*, Sen puts human freedom in the centre of economic development. In his view, freedom, justice and democracy are the absolute prerequisites for making any economic development and the fight against poverty possible in the first place. Where freedom is restricted in favour of economic development, for instance, in today's China, political and social contradictions increase, and economic development will eventually reach its limits. The political answer to the resource curse is, therefore, the strengthening of freedom and democracy.

Chapter 7

Ways Out of Dependence: Solar or Nuclear?

The International Energy Agency writes in its *World Energy Outlook 2006*, that the continuation of the current energy-political trend is expensive, harms the environment and threatens security. It criticizes the dependence on fossil fuels, the growing market power of a few energy-exporting countries, and the risk of climate change. However, what are the alternatives: Is the world ripe for a rediscovery of nuclear energy? Or can renewable energy sources, such as the sun, wind, water and biomass, satisfy the world's growing hunger for energy?

Escaping into nuclear energy

For a few years, nuclear energy has experienced a new boom. According to the International Atomic Energy Agency (IAEA), 30 countries operated 443 commercial reactors at the end of 2005. This covered only 5 per cent of worldwide energy consumption, but still 16 per cent of the electricity used. The large majority of the nuclear power stations are operated in the traditional industrialized countries (USA 104, France 59, Japan 55, Russia 31, the UK 23, Germany 18). South Korea has currently 20 nuclear power stations, India 15 and China 9. The greatest growth potential for the use of nuclear energy thus lies in the countries of East and South-East Asia. The IAEA estimates that, if it receives the necessary support, worldwide nuclear capacity might quadruple by 2050 'in view of a growing world population, a backlog demand in developing countries and the necessity of climate protection'.

Initially, it was the fear of climate change that made some governments turn once again to the CO_2-free energy resource nuclear power. However, strictly speaking it is untrue that no CO_2 is produced in the generation of nuclear power. The production of uranium, the fuel still used for operating most nuclear power stations, is extremely energy intensive. Mining machines operated with oil are normally used. The construction of the power stations as well as the transport and storage of nuclear waste also influence the energy balance of nuclear power negatively. The energy input necessary to construct plants for the production of renewable energies is significantly lower.

However, the energy security argument has gained in importance over the past few years. Besides, prices for fossil fuels have risen dramatically and threaten to increase even further in future. Expensive nuclear energy has thus become more economical than it was in the 1980s and 1990s when oil prices were low. In principle, electricity generated from nuclear power can replace electricity generated from fossil fuels such as coal, oil and gas. Nuclear power stations can also produce thermal energy, that is heat, for operating industrial plants and for heating purposes. Over the long term, it is even speculated that nuclear energy could be used for hydrogen hydrolysis and thus for generating hydrogen fuel. But whether hydrogen technology will overcome the existing technical obstacles and thus produce the fuel of the future is still written in the stars.

The fear of energy-political dependence that many countries have must be taken seriously. It is primarily the states of Eastern Europe as well as East and South-East Asia who are focusing on the extension of nuclear energy to reduce their dependence on imports of fossil fuels. Following the Chernobyl reactor accident and the changes in the Soviet bloc that began at the same time, most nuclear programmes in Central and Eastern Europe came to a halt. Today, more often than not, the same technocrats have once again secured their positions in the energy ministries and pursue their old projects. The most important argument in favour of the proliferation of national nuclear programmes, which is brought up repeatedly, is independence of energy imports from Russia.

The three Baltic states, together with Poland, are planning a joint nuclear power station that will allow them to become less dependent on Russian gas imports. The new Baltic nuclear power station is to be erected at Ignalina in Lithuania, on the site of a former Soviet reactor that had to be shut down at the demand of the EU. In truth, the Baltic states could alternatively be supplied with electricity and gas from other EU countries if it was not for the inadequate extension of the trans-European energy network. In such a uniform European supply region, electricity would come from different sources, procuring gas not only from Russia, but also from Norway or North Africa. The erection of transnational pipeline systems via Poland into the Baltic region would certainly be feasible within a shorter time and at lower cost than the construction of a new nuclear power station at Ignalina. In fact though, the project might fail because energy policy cooperation between Poland and its Baltic neighbours, who all would have to cooperate to make the new reactor an economic reality, is as weak as between Eastern and Western Europe in general.

Other Eastern European countries also wish to keep to nuclear power to avoid becoming dependent on energy imports from Russia. Bulgaria, which undertook to shut down its only nuclear power station after joining the EU, now wishes to reverse that decision. In the meantime, a protest movement has collected several hundred thousand signatures demanding that the Bulgarian nuclear power station in Kosloduj should be retained. Armenia, whose electricity generation is sourced almost completely from the only nuclear power station in the country, wishes to replace its accident-prone reactor with a completely new construction on the same site. The problem in this regard is that Armenia's nuclear power stations were built on an earthquake-prone geological fault, with no suitable alternative locations available in the small country.

Finland, the only Western European country that is currently building a new nuclear power station, also cites its high dependence on Russian imports as justification for the project. As all these examples prove, most countries do not use economic or technological arguments in favour of keeping or extending nuclear energy, but rather the wish for political independence. If Europeans can succeed in placing their relationship with Russia on a more reliable footing, while at the same time reducing their economic dependence on natural gas through the use of renewable energy and by implementing a sensible strategy of diversification, an important argument in favour of nuclear energy would fall away.

More significant than the relatively small-scale nuclear programmes in European states are the plans of some major emerging economies. India, China and Brazil already decided on their nuclear programmes decades ago, but did not implement them fully, so that the percentage of nuclear power in their power generation is comparatively low. Nuclear power in China, for instance, is currently responsible for only 1 per cent of the basic energy demand. However, China does not want to depend on oil and gas imports alone anymore, and plans to at least quadruple the contribution of nuclear power. India also focuses on the erection of nuclear power stations to reduce its import dependence. The extension of the Indian nuclear programme came to a halt because the West imposed an embargo on modern nuclear technology and uranium following India's nuclear weapon tests in the late 1990s. The embargo is to be lifted now on the initiative of the US. The US government cites the possible important contribution of nuclear energy to the development of the Indian economy, climate protection and certainly also to the cash boxes of American exports. It appears that the continuation of the Indian nuclear weapon programme and thus the danger of a nuclear race in the region are tacitly accepted. However, a US–Indian nuclear cooperation treaty, that would have allowed the sale of fissile material to India, has not yet been ratified by the Indian parliament.

At the end of 2006, the Chinese government announced that it would extend its nuclear-political cooperation with India's neighbour and rival, Pakistan. Two great powers, the US and China, would thus be behind the regional nuclear race between India and Pakistan.

South Korea, the world's second largest coal and third largest oil importer, also intends to build additional nuclear power stations to satisfy its hunger for energy. Four additional light-water reactors are to increase South Korea's nuclear power percentage from 40 to 60 per cent by 2017. South Korea has a substantial anti-nuclear movement that fears that one of South Korea's nuclear power stations might become the target of a missile attack from the

Communist North. But since the former nuclear-critical democratic movement of the country has replaced the old military dictatorship, the quest for energy-political independence from such powerful neighbours as Russia and China is more important than the fear of a nuclear worst-case scenario. South Korea, and also the island state of Taiwan, are mainly concerned about an interruption in the oil and gas import routes. Both countries are 100 per cent dependent on oil imports from the Near East, with imports almost exclusively being shipped by sea. In the event of any conflict between China and Taiwan or, alternatively, the US, the tanker routes along the Chinese coast could be blocked or interrupted by wartime activities.

Japan, on the other hand, announced after the Kyoto Conference that it would achieve its CO_2-saving target by erecting 20 new nuclear power stations. In the meantime, these plans have been revised. Current plans only – but still – provide for 12 new reactors.

In many developing countries, possessing nuclear energy is still regarded as the pride of technical progress and a matter of national prestige. However, states that rely on nuclear energy place themselves in a precarious political and economic dependence situation. As is the case with oil and gas, nuclear power is not a local energy resource. At present only a few countries and companies have the capacity to erect nuclear power stations and maintain them properly. Currently, nuclear exporters Russia, Japan, the US and France are competing for the market that is developing in East and South-East Asia. Besides, international trade with nuclear technology is subject to strict regulations. As the example of India shows, an embargo on sensitive technology or the nuclear fuel uranium may halt the nuclear programme of one of the most populous countries in the world. Countries like Iran that try to obtain modern nuclear energy in a roundabout way must accept year-long delays in their plans. In the case of Iran, there might also be international sanctions that go far beyond the energy sector.

Uranium, the fuel used for conventional nuclear power stations, is also a diminishing resource. Only very few countries have their own uranium reserves. Even though uranium can be stored over long periods, the reserves that are currently known will only last for a few more decades. In 2004, 40,000 tons of uranium was extracted from mines, while consumption amounted to 68,000 tons. At present, the balance comes from military stocks, which, however, will be used up at some point. By nuclear reprocessing or so-called ‘fast breeder reactors’, it is possible to recycle and reuse nuclear fuel rods. However, this process produces materials that could be used for nuclear arms, and the proliferation risk increases. Besides, the fast breeder reactor technology is difficult to control. For this reason, Germany shut down its model reactor in Kalkar in North Rhine-Westphalia after operating it for only a couple of hours.

Today Canada and Australia are the most important exporters of uranium. Both countries have large unexploited reserves. Western countries often argue that the supply of uranium from Australia and Canada, both stable democracies, is more reliable than oil and gas imports from crisis-prone regions such as the Near East and Africa. Nevertheless, the energy industry cannot rely on the planned extension of uranium mining. There are already massive protests against uranium mining due to its extremely high environmental impact. In Australia, for instance, a controversial debate is ongoing on whether it would really be in the country's interest to extend harmful uranium mining to enable China to use nuclear energy on a large scale. Besides, numerous uranium reserves are located beneath the holy places of the Australian Aboriginal inhabitants. If Australia takes its democratic principles and the protection of its minorities seriously, these reserves will not be exploited.

Other uranium mines can be found in Kazakhstan, Uzbekistan, Russia, Niger, South Africa, Namibia and Iran. Minor reserves, which, however, it would be uneconomical to exploit, are dispersed throughout the world. One example is the uranium production of the Wismut Mining Company in eastern Germany, which was stopped after reunification. Wismut's mining operations produced a moon landscape with mountains of radioactive waste. The clean-up activities that followed proved to be far more expensive than the economic gain from the decades before. In general, the heavy metal uranium is extremely rare. It is therefore unlikely that any further major deposits will be found.

The biggest problem of East Asian countries that today focus on nuclear energy is neither access to the technology nor the procurement of uranium, but rather the disposal of increasing quantities of nuclear waste. Since the 1980s, South Korea, the country with the largest nuclear programme in the region, has been looking for a suitable repository for the 6500 tons of radioactive waste that has accumulated. Taiwan tried to export its nuclear waste to North Korea or the Marshall Islands. The North Korea project was abandoned following protests of the South Koreans. The 100,000 barrels containing Taiwanese nuclear waste, which are currently stored on the Marshall Islands, must be returned to Taiwan by 2013, because of protests by the local population.

The counter-arguments used by the anti-nuclear power station movement in the 1970s and 1980s and which resulted in an actual nuclear moratorium in the industrialized states of the West, continue to be valid today. The allegedly safe new reactor types, which the nuclear industry has been promising for years, have not yet passed practical tests. Furthermore, people did not believe that the nuclear accident at Three Mile Island or the catastrophe of Chernobyl were technically possible. In summer 2006, a core meltdown and thus a repeat of the worst-case scenario of Chernobyl almost happened in Sweden, one of the countries that experts believed had particularly high safety standards. Several safety systems and emergency power generators in the Forsmark nuclear power station failed at the same time; there was the real threat that the cooling of the fuel rods might fail. Forsmark shows that even the most sophisticated technology can fail.

It is operated by humans, is subject to political and economical interests, and may be damaged or destroyed in crises, wars or during natural disasters.

Siamese twins

Since the nuclear-political debate of the 1980s, there have also been a couple of significant additional arguments. Three of today's most important challenges in terms of security are international terrorism, the proliferation of weapons of mass destruction, and the break-up of states into ungovernable units, or shadow states, which are ruled by criminal warlords. All three developments make nuclear power even more prone to military or terrorist misuse than before.

Civil and military uses of nuclear energy have been Siamese twins from the beginning. Nuclear technology, expertise and material, can be transferred. Nuclear experts may leave their home country or sell their knowledge to criminal networks. As early as 1975, a CIA study warned against the danger of nuclear terrorism. The Nuclear Non-Proliferation Treaty was drawn up in the 1960s, motivated by the fear that there would soon be close to 30 countries with nuclear weapons because of the proliferation of civil nuclear technology. At the time, numerous large developing countries, some of them military dictatorships such as Argentina and Brazil, tried to obtain the bomb. South Africa admitted later that it had developed the atomic bomb. With it, the apartheid system wished to ensure its survival and prepare itself for a possible military attack by its neighbouring African states. Iran's nuclear ambitions also began at that time. Until the end of the Cold War, the Non-Proliferation Treaty and the control that the superpowers exerted with their nuclear technological expertise over their respective power blocs, and almost all the other countries, prevented this horror scenario. Today, the hitherto successful politics of containment to prevent nuclear proliferation is threatening to collapse. The most important reason for this is the renaissance of peaceful industrial use of nuclear energy, with the related expansion of technology that can also be used for military purposes. Besides, the US, through its policy towards Iran, has been undermining the successful Non-Proliferation Treaty. In this respect the US has opted for the use of military and economic threats rather than inspections by the International Atomic Energy Organization (IAEA), which proved very successful in the case of Iraq.

After the first Gulf War in 1991, inspectors of the IAEA and from the US discovered that Iraq had a rather advanced military nuclear programme and was on the verge of building its first bomb. In 1998, India and Pakistan surprised Western secret services when they detonated atomic bombs. International sanctions were imposed on both countries for a certain period. However, those on India will be discontinued soon as part of a far-reaching nuclear cooperation agreement with the US. Both North Korea and Iran have also shown how difficult it is to distinguish between civil and secret military nuclear programmes. Often, the same facilities and technologies are used to both ends. At a time of international terrorism, collapsing states and the simultaneous increase in the number of states that are capable of the civil application of nuclear energy, the danger of proliferation of the use of nuclear material for military and criminal purposes is also increasing.

All the new nuclear powers started their military programmes through, or at least hid them behind, civil nuclear programmes. India produced the plutonium for its first atomic explosion in 1974 using a research reactor designed in Canada. A research reactor, this time of Russian origin, was also the beginning of the North Korean programme. In addition, South Africa, which discontinued its nuclear weapon programme following the end of apartheid, has extended its originally civil programme to military applications. Pakistan succeeded in stealing the technology for its centrifuge programme for the enrichment of uranium from the Netherlands. Indirectly, Pakistan's programme is based on the civil use of nuclear energy in another country.

Research reactors operating in numerous countries, including Germany, often use highly enriched and thus weapons-grade uranium. The 'atomic egg' in Garching, the first research reactor in Germany, also used highly enriched uranium (HEU). Originally, it was erected on the initiative of the then Minister of Defence, Franz-Josef Strauss, with the ulterior motive of keeping the option of nuclear weapons open for Germany as well.

Nuclear recycling plants separate plutonium, another weapons-grade material, from the burnt nuclear fuel rods. Uranium enrichment plants, which produce the material for the fuel rods from natural uranium, can also be used for military purposes. In the case of Iran, the international community fears that the centrifugal nuclear enrichment systems are not only used for producing the lower degree of enrichment necessary for nuclear fuel rods but also a significantly higher level of enrichment that is required for producing nuclear explosives. Throughout the whole nuclear fuel cycle there are dangers lurking, such as the difficulty of distinguishing between civil and military use, and the possible sidelining of dangerous material.

Besides, there is the risk that civil nuclear plants might become the targets of military attacks. Be it the two Koreas, Taiwan and the People's Republic of China, or India and Pakistan – a military conflict in one of these tension areas could rapidly escalate to a nuclear confrontation. Civil nuclear plants can be attacked from the air or from the ground, or might be hit by accident. During the Iran–Iraq war, the Iraqi air force repeatedly bombed the construction site of the Iranian nuclear power station in Bushehr. Iraq's own Osirak reactor was destroyed by Israeli air attacks even before completion. To date, there have not yet been attacks on any reactors containing nuclear fuel rods, but this cannot be excluded.

Following 11 September 2001, the German Commission for Reactor Safety investigated, by order of the Federal Government, whether the steel-reinforced concrete structures of German nuclear power stations would provide sufficient protection against the targeted crash of a passenger aircraft. There are reasonable doubts, at least in the case of older plants like the power station in Obrigheim, Baden-Württemberg. The protection structure of German power stations has been designed to withstand the crash of a sporting aeroplane or a military aircraft, but not that of a passenger aircraft the size of an Airbus 320. American reactors are not secured against aircraft crashes at all, and the protection of French reactors is considerably poorer than that of German reactors. Most Russian nuclear power stations do not even have a protective structure to limit any explosion to the interior of the reactor and prevent radioactivity escaping.

In times of war, control of nuclear material might also be lost. When the Americans invaded Iraq in 2003, they forgot to secure the country's most important nuclear research plant against looting. Afterwards, the IAEA determined that documents as well as nuclear material had disappeared. Some of the material was rediscovered on the black market, but only a minor part of it could be seized and confiscated.

In times of international terrorism, there is a growing danger that nuclear material might fall into the hands of terrorists, or even common criminals who might sell it on to terrorists. During searches of Al Qaeda training camps in Afghanistan, sketches for building basic nuclear weapons were found. Another fear is the production of so-called 'dirty nuclear bombs' by terrorists. This is an explosive device consisting of conventional explosives with added radioactive material to be distributed as widely as possible in an explosion. A nuclear fission and thus nuclear explosion does not take place. In its major report on its weapons of mass destruction to the UN in December 2002, Iraq admitted that it had temporarily investigated the feasibility of such a weapon.

Basic nuclear weapons are not very difficult to build. The instructions can be found on the Internet. Even the carrier rockets for transporting an atomic bomb safely to its target are increasingly easy to obtain. As the latest conflict has shown, the rockets of Hezbollah from Southern Lebanon now reach far into Israel. It is only a question of time when rockets of Iranian or North Korean origin with even greater ranges will be available on the international market. Suicide bombers would not even need such carrier weapons. They could blow themselves up with the bomb on the spot.

What has so far made it difficult or impossible for terrorists to produce atomic weapons was access to the necessary nuclear material, in particular, highly enriched uranium. The required technology is difficult to come by, and can only be operated by highly specialized experts. For carrying out such attacks, terrorists would need the backing or at least silent connivance of a state that either has nuclear weapons or nuclear weapons-grade material. Iran is among the group of states that presently support terrorist organizations like Hezbollah. Pakistan's secret service has been supplying the Taliban in Afghanistan with weapons for years. Even an emerging market like Iran that has pursued a military nuclear programme for decades is not yet capable of enriching a sufficient amount of uranium for building its first bomb. However, this is exactly the danger of a massive extension of civil nuclear power. The total amount of nuclear material in circulation worldwide would increase dramatically. Even though it is currently unlikely that nuclear technology of states could fall into the wrong hands within the framework of such cooperation, this cannot be excluded in the future.

Besides, terrorists would not even have to own the bomb themselves to trigger a nuclear catastrophe. The same effect could be achieved by a targeted attack on a nuclear power station, on other often little secured civil nuclear research plants, or on transported nuclear material. The effect of quantity would come up once again. The more nuclear power stations a country has, the more material will be transported by land, sea and air. The risk of accidents, as well as the risk that nuclear transports might accidentally or intentionally be targeted in warlike operations, will increase proportionally.

Some countries with weak government structures and accordingly rudimentary security provisions are also interested in nuclear power. How certain can we actually be that Pakistan, a country that owns not only a civil but also a military nuclear programme, will continue in its current form for another 20 years and will not break down into radical Islamic subunits at some stage? Is it possible that the Chinese centralized state might disintegrate at some time, as happened to the Soviet Union in the early 1990s? In any case, everybody who advocates the further extension of nuclear energy should think in terms of these historic time frames. After the disintegration of the Soviet Union and the collapse of the gigantic nuclear industry of the former great power, it was no longer possible to determine the whereabouts of substantial quantities of nuclear material. Not only radioactive substances, but also sensitive nuclear technology, and possibly a few tactical nuclear warheads, might have disappeared in the meantime. Particularly long-term and dangerous effects were caused by the entry of thousands of well-trained Soviet nuclear experts onto the international labour market. Many of them found work in Pakistan and in the Arabian Gulf states. Today, more than 1000 Russian experts and engineers work on the construction site of the Iranian nuclear reactor in Bushehr.

In the end, politicians must weigh up the dangers of nuclear power against other risks, such as global climate change and the dependence on fossil fuels, as well as against developing alternatives such as renewable energies. The German government's scientific advisory committee for global environmental changes writes in this regard:

> If the disastrous potential cannot be reduced in a sensible manner or only at exorbitant cost even with the best of efforts, then ... such

> source of risk should only be authorized if, on the one hand, the benefit of this risk source is existentially significant and, on the other hand, it can be ensured that all technical, institutional and organizational options are used so that the disaster will not happen in the first place... This second prerequisite is particularly important if such risk sources are exported abroad within the framework of technological transfer.

The question whether these prerequisites are fulfilled under the conditions of collapsing states, international tensions and major ideological conflicts, such as the one between the West and radical Islamism, can confidently be answered with a 'no'. 'The large-scale worldwide use of nuclear energy', as physicist and philosopher Carl-Friedrich von Weizsäcker said, 'demands in consequence a worldwide radical change in the political structure of all of today's cultures. It demands that the political institution of war, which has existed at least since the beginning of civilization, must be overcome.' However, as this is a Utopian dream, the existing risk must be restricted and reduced as far as possible through international conventions and institutions.

However, not only environmental politicians doubt that the use of nuclear energy is necessary and indispensable for securing our energy future. After all, there is the alternative of the huge potential of renewable energies. This hypothesis is investigated below.

Energy security through renewable energy sources

Anyone who does not wish to use nuclear energy must focus on renewable energy. A national energy policy that does not depend on fossil fuel imports but on the use of renewable energy resources could also contribute to peace. However, can renewable energy really fill the gap created by renouncing the use of nuclear power and the decrease in fossil reserves? In particular, for developing and emerging countries, it is important that the potential of the sun, wind and water for energy supply is not just proven theoretically. Access for these countries to technologies and capital must also be ensured. After all, renewable energy has to prevail over diehard interests of the oil, gas, coal and nuclear lobby.

Efficient energy use and the transition to renewable energy do not only increase our safety, they also reduce the dependence on expensive and insecure energy imports. At the same time, an energy policy that focuses on substituting fossil fuel resources contributes to climate protection. Not only environmental politicians but also security experts argue that climate protection is active long-term security policy. Former US Secretary of State Madeleine Albright realized in the late 1990s that climate protection was one of the most important foreign policy challenges faced by the US. She added: 'Nowhere else have I experienced negotiations as complex as these.' In his keynote address on environmental policy, former British Prime Minister Tony Blair stated: 'We will not be able to live safe if climate change destroys our means of existence.'

The theoretical potential of renewable energy is indisputably high. According to estimates of the International Energy Agency and others, renewable energy could replace the current and future energy consumption completely and would even permit a three- to sixfold increase in worldwide energy consumption. The theoretical potential of renewable energy is thus definitely higher than that of nuclear power, the increase of which is restricted by declining natural uranium reserves, even if fast breeder reactors are used and fuel is recycled.

For the time being, the use of this potential is restricted by the current basic economic conditions. The costs of generating electricity and heat from solar, wind and geothermal energy are still higher than the costs of using coal, oil and gas. However, if prices for fossil fuels increase and the costs for renewable energy technologies decrease, this price gap will close over the next few years. State incentives such as the introduction of CO_2 taxes and emission certificates could accelerate this process. Most major energy companies have invested a lot in their fossil-fuel-based infrastructure. If politicians wish to help leveraging new environmentally friendly technologies, they must make use of economic incentives and political pressure to ensure that not only small pioneer companies but also the large energy companies change direction. An approach that is merely based on economic aspects is in any case not helpful. If it is true that the dependence of our economies on oil and gas imports from crisis regions and authoritarian states threatens our democracy and security, the transition to renewable energy takes on a strategic security aspect. Our own security should be worth the financial investment that goes beyond a mere economic approach.

Apart from economic obstacles, political resistance against renewable energy installations must also be overcome. Both nature conservation regulations and the interests of the local population must be considered when building new wind and hydropower plants. This applies above all to the construction of large dams that often require massive resettlement exercises. An example is the Three Gorges Dam on the Chinese Yangtze River. It was commissioned on 20 May 2006 and generates, under normal circumstances, an electricity equivalent to 14 per cent of total German consumption. Undoubtedly, the dam contributes to avoiding CO_2 and to climate protection, as China would otherwise have had to erect further coal and nuclear power plants. For this reason, even the German Ministry for the Environment gave its approval when it was approached to authorize the export of German turbine technology for the gigantic dam. However, the negative side of the project was that 24,000 hectares of land had to be flooded and almost two million people had to be resettled. Consequently, it is imperative that renewable energy projects must be carried out following environmental impact assessments and with the involvement of the local population. The Three Gorges Dam project would only have been genuinely environmentally friendly if the necessary energy had been generated at a different location using smaller renewable energy systems or had been obtained through efficiency measures. However, in authoritarian states like China, the population is not involved in energy-political decisions. Consequently, it is even more important that a country like Germany, when granting export permits and loans, makes sure that certain minimum standards in terms of public involvement and environmental protection are also observed abroad.

The potential for renewable energy varies from region to region. Large water power plants such as the Three Gorges Dam that secure the electricity supply of a metropolis or an entire industrial region in one fell swoop can only be erected on a few major rivers. On the other hand, small hydropower plants suffice to fulfil the local electricity need in many sparsely populated regions, in particular, in mountainous regions. Wind power can be utilized primarily near the coast. The use of solar energy is economical mainly in southern countries. Solar installations require a large space. But if we look at all renewable energy resources available, almost every country in the world has significant potential. Even in populous, densely populated and economically strong emerging markets like India and China, the use of biomass has a high potential.

Some countries, for example, the industrialized nations in Europe or the US, will cover part of their energy requirement through imports in future as well. The sunny continents of Africa and Latin America, on the other hand, do have the potential of exporting electricity generated from solar power or fuel from biomass. Biofuels can replace fossil fuels for cars, for heating and for heat and electricity generation in power plants. They are already being added to fossil fuels today. What is important, however, is that a global economic system based on renewable energy is structured so as to avoid old dependencies on oil and gas being replaced by new unilateral dependencies.

Will the age of oil be followed by the hydrogen era?

Since oil prices have risen dramatically, even oil and automotive companies have discovered renewable energy. Technological visionaries regard hydrogen as the energy resource of the future. Will the oil era thus be followed by the hydrogen era?

Hydrogen is the most abundant element on our planet. As hydrogen is combustible and can be liquefied under high pressure, this ubiquitous element is basically suitable for both electricity generation and as fuel. A possible application that fires the imagination is the electrolytic disintegration of water into hydrogen and its further conversion in fuel cells into electricity. This could not only drive car engines, but also heat and industrial plants. Jeremy Rifkin, one of the prophets of the hydrogen era, even imagines that in future the toaster at home and the mobile phone when travelling will function on fuel cells. For now, however, security problems and technical hurdles stand in the way of a large-scale entry into the hydrogen era. If large quantities of hydrogen leak, it reacts with the oxygen in the air and is converted to water. This spontaneous reaction is called the oxyhydrogen effect that every schoolchild knows from chemistry lessons. What is very funny in seventh grade represents a considerable stumbling block at industrial level.

Because hydrogen is the smallest element in the periodic table, it is difficult to store and to transport. Only special alloys and composite materials expensive to manufacture are suitable in this regard. Normal natural gas pipelines and tanks cannot be used. Hydrogen filling stations would have to be specially built from scratch. The technical problems of the construction of a wide hydrogen infrastructure also have political effects. Due to the highly complex technology and the strict safety standards that would have to be observed, hydrogen technology would only be a real option for developed industrialized countries and states with stable governments. In this regard, the hydrogen industry is similar to the nuclear industry.

Other renewable energy technologies, photovoltaic cells or solar hot water systems have often been advanced by creative and meticulous individual inventors. Every step of the manufacture, transport and use of hydrogen requires a highly complicated technology. The entire technological chain could be patented. Only a few large companies would be able to take on this capital-intensive task, and would then control this technology.

Besides, the type of primary energy resource that will be used for the electrolysis from water to hydrogen will also be decisive for the environmental and climate-friendliness of hydrogen as an energy resource. On principle, it can be done using electricity from fossil power plants, from renewable energy such as waterpower or photovoltaic cells, and also from nuclear power stations. It has therefore been repeatedly suggested to generate hydrogen using solar energy in the Sahara desert and to transport it via pipeline to Europe. The EU has put this plan on hold for the time being because the pipeline technology has not been developed sufficiently to accommodate the above-mentioned volatility of hydrogen. Canadian environmental politician Maurice Strong, who organized the World Summit in Rio in 1992 and was subsequently appointed as CEO of the hydropower group Quebec Hydro, has suggested that hydrogen should be generated using the electricity from large dams in Canada and exported by tanker to the US and across the Atlantic. Under President Bush, the US energy authority investigated the possibility of producing hydrogen for America's car fleet using nuclear power, which would significantly increase the above-mentioned security and proliferation problems.

The biomass boom – joint energy and agricultural reform?

Eastern Europe is only one of the regions where utilization of biomass might be an important part of the energy future. In Russia, Belarus and Ukraine, traditionally the breadbaskets of Europe, the population is decreasing because of emigration and low birthrates. At the same time, large agricultural areas that were inefficiently used during the Soviet era are available. The cultivation of plants for use as biomass could open up new perspectives on agriculture in Eastern Europe. The necessary reform of their agricultural policy could thus become an important element of an energy turnaround for Eastern Europeans.

Major German agricultural companies as well as the petrochemical and automotive industries have, for a couple of years, been contemplating signing contracts with Ukraine in respect of large tracts of arable land for producing biodiesel and ethanol as petrol replacements. In principle, biogas could replace natural gas imported from Russia. Apart from saving fossil fuels, this would also improve the poor air quality in Eastern Europe's main cities.

However, bioenergy would have to be produced without causing any new inherently negative environmental impact. Large amounts of fossil fuels in the form of fuel for the harvesting machines, processing, chemical fertilizers and pesticides are often used for cultivating grain, rapeseed or other crops for generating biomass. If very intensive agriculture is practised, the energy balance might even be negative.

The agricultural industry is also planning to cultivate genetically optimized species on a large scale for use as biomass. The vision is that genetically engineered energy plants would provide a higher energy yield, would require less fertilizer and pesticides and would not be subject to the strict EU food laws. The cultivation of energy plants is considered one of the major areas of application of green genetics. However, it is hardly possible to separate cultivation areas for energy plants on the one hand and food and fodder plants on the other. Even hybrid cultivars and genetically sterile energy plants mutated in field tests and disseminated on neighbouring fields. In a free market economy and with fluctuating prices, farmers would regularly alternate between cultivation for energy and for food production. Currently, public opinion critical of genetic engineering prevents large-scale entry into this agricultural technology in Western Europe. In Eastern Europe, on the other hand, in particular, in countries like Ukraine that are not bound by the relatively strict EU laws, there is the danger that Western funds could be used for cultivating genetically engineered energy plants for export before these countries get the chance to decide for or against this route after informed social debate. The question of future energy policy is thus once again a question of democracy.

In numerous countries of the South, energy supply is traditionally based on biomass. This includes the burning of wood, straw or manure. This traditional use is often hardly environmentally friendly, leads to poor air quality in the cities and results in the destruction of natural resources. The International Energy Agency (IEA) reckons that today 70 per cent of the energy from renewable resources comes from the traditional use of biomass. However, a large part of this use is not sustainable and, in some cases, for instance in the Sahel region, results in massive conflicts over land usage, the loss of arable areas, displacement and political tensions.

An example of modern biomass use is China. China is already the world's third largest producer of ethanol. At the end of 2005, 2.5 per cent of Chinese fuel demand was covered by ethanol produced from grain. By 2020, national production is to quadruple. Even if other fast-growing agricultural crops and agricultural waste are used for biomass production, a maximum limit for the bioenergy potential is foreseeable in a country like China where food for its growing population is already being imported. China, India and other populous growth economies in the South will only be able to fulfil their fuel demand from renewable resources if they import biofuel or hydrogen instead of crude oil. The same applies to North America and Western Europe. Even if the importers remain the same, the exporters will change.

Another decisive issue is that food production and the cultivation of energy crops must not be mutually exclusive. In the end, farmers in developing countries will grow the crop that will earn the highest price on the market. Basic local state regulations must therefore ensure that people will not starve while their food is used in a tank. Lester Brown, the founder of the Washington-based World Watch Institute, expressed it as follows: 'Then, the supermarket will compete with the filling station. Put differently, it would be a struggle between 800 million car owners worldwide and two billion people who simply wish to survive.'

Renewable energy and EU climate policy

The European Union has undertaken to increase the percentage of renewable energy in general energy consumption to 20 per cent by 2020. This general target will be supplemented by sectoral targets for biofuels, the electricity sectors and the application of renewables to produce heat in the building sector. This would not only achieve the EU climate protection goal, but also reduce import dependence. Most advancement has been made in countries like Germany, Spain and Denmark, which have leveraged wind power and other new technologies through feed-in regulations for renewable energy and other price incentive systems. Other Western European member states such as France and the UK still have a lot of catching up to do. The largest potential for an extension of renewable energies exists in the new member states. The heritage of state-directed economies in Eastern Europe, low energy prices and a major-projects-mindset have prevented the transition to a decentralized, sustainable energy supply. This is despite the fact that Eastern European countries would have a major advantage in freeing themselves from their dependence on energy imports from Russia.

To date, hydropower is mainly used. However, the erection of additional large-scale power plants clashes with concern for nature conservation. Besides, the Baltic states and Poland have favourable conditions for an extensive use of wind power along their coasts. According to studies, Estonia could fulfil its electricity demand entirely from wind power. This would eliminate the dependence on imports from Russia while the construction of the planned new Baltic nuclear power station would no longer be necessary. To date, biomass has mainly been used in the traditional way, for example, in wood heating for residential homes in Lithuania. But the systematic cultivation of energy plants in countries with a strong agricultural structure like Poland would have significant potential. Modernizing Polish agriculture and its energy supply would go hand in hand. In Hungary, geothermal energy is also used for generating hot water. Primarily in sparsely populated areas in Eastern Europe and in the east of Russia that are not connected to the trans regional energy networks, geothermal power plants are competitive even for electricity generation. For instance, a geothermal power plant is being constructed with assistance from Germany on the peninsula of Kamchatka in eastern Siberia.

Even though all Eastern European countries except Russia are net energy importers, the promotion of renewable energy does not yet have the necessary political priority in these new EU countries. The legal prerequisites for promoting renewable energy have, in the meantime, been created in most countries in Central and Eastern Europe, yet implementation is something else. In most Eastern European countries, energy prices are still subsidized. The introduction of prices in line with the market would make it possible for renewable energy to compete with Polish coal, and oil and gas imported from Russia. The price increase for Russian natural gas is therefore positive for developing a sustainable energy supply. However, sudden overnight price increases resulted in a situation where the limit of what is socially acceptable to many poor energy consumers was exceeded.

Nevertheless, the EU has a competitive advantage over other players on the global market as far as renewable energy is concerned. Germany is the world market leader in wind power and shares the top position in photovoltaics with Japan. Sweden and Iceland intend to become completely independent from energy imports within a few years. According to Germany's green environmental politician, Reinhard Loske, Germany could reduce its CO_2 emissions to zero by 2050 by promoting renewable energy and efficiency measures. Using a mixture of user-orientated research, favourable market mechanisms such as the German feed-in legislation and the large number of small and medium-sized companies, together with an immense growth market in Central and Eastern Europe, Europe's renewable energy industry could reach the top of the world market. The EU could thus also regain leeway with regard to countries like Russia and the Near East that continue focusing on fossil fuels.

A solar partnership with North Africa

Europe and the countries in North Africa depend on each other. Apart from Russia, the EU states import their oil and gas primarily from Maghreb states such as Libya and Algeria. Egypt, the most populous country in North Africa, is also entering the European gas market on a massive scale. On the other hand, most imports to the Mediterranean rim states in North Africa come from the EU. Apart from sharing historical experiences, such as colonialism and close cultural connections between the two regions, North Africa and Europe are also connected by a number of common problems. This includes demographic developments, migration, organized crime and transnational terrorism. Mutual dependence of import and export countries in the region is an important argument for closer political cooperation and economic integration.

Countries such as Libya and Algeria earn a major part of their economic output and their state budget from energy export. This economic structure that is unilaterally orientated towards the energy industry holds two dangers. First, other industrial sectors were neglected in the past. Second, it is completely unclear what is going to happen once the oil and gas wells of the countries in North Africa run dry. As the time when oil and gas reserves will decline is foreseeable, a change in direction towards renewable energy should be undertaken now. Two prevailing conditions in the countries of North Africa can make them the optimal locations for the production of electricity from solar energy. Statistically, the sun shines on average for more than 300 days a year. Also, the infertile desert or semi-desert areas would offer sufficient space for building large photovoltaic systems. In the end, the countries in the southern Mediterranean will depend on permanently available renewable energy for their own economic development. Moreover, energy – in particular electricity – could then also be exported to Europe in future. In order to export the generated electricity directly, a new efficient high-voltage network would have to be erected to connect North Africa with highly populous centres in Europe. Such a 'trans-Mediterranean' electricity network would be a logical extension of trans-European energy networks as already being promoted by the EU Commission.

In many countries in the Mediterranean, energy poverty is a significant obstacle to economic development. However, sustainable economic development is the basic prerequisite for mitigating social conflicts and creating political stability. Nowhere else is this more obvious than in the densely populated area of the Gaza Strip under Palestinian autonomy. The Jordan Academy of Sciences recently suggested that electricity supply for the population and the economy of the Gaza Strip should be secured by solar power from the Egyptian peninsula of Sinai. The required areas are not available on Palestinian territory. A secured energy supply organized in regional agreements for the Palestinian territories would also promote peace. Neither Israel nor Egypt can be interested in a collapse of Palestine's economy, which would escalate the political situation and force young Palestinians to stream into neighbouring countries as economic refugees.

Solar energy can also help alleviate water poverty in the region. A shortage of water resources combined with a rapidly increasing population probably holds even more social explosiveness than energy scarcity. In Yemen, a country where the population doubles every 25 years, not one river reaches the sea today. All the water resources are 100 per cent utilized. The water table beneath the capital Sana'a is dropping by 7m per annum. Violent struggles over water rights and access to wells among the tribes in Yemen are already on the increase. Yemen caught the attention of international media with a series of spectacular kidnappings of Western travellers, many of them aimed at extorting political concessions from the Yemenese government, for example, access to energy and water for tribes in remote regions.

However, it is not only in Yemen that the mixture of demographic explosion and scarcity of resources has reached a critical point. If Egypt's population continues to grow at today's rates, the country will require a 'second Nile' by 2050. Egypt is the most populous Arab country and currently one of the few stable factors in the region. The son of President Mubarak, who has ambitions to succeed his father in the highest office of the state, announced in his maiden political speech in summer 2006 that he planned to solve the energy problems of the country by erecting nuclear power stations. These would also power seawater desalination systems and thus alleviate the threat of water scarcity. However, if Egypt were to become the second country in the region after Iran to use nuclear power, this would also result in the threat of a further extension of military nuclear technology in the Near and Middle East. It would therefore be much more sensible to operate the necessary seawater desalination systems using solar energy, while at the same time tackling the energy and water scarcity prevalent in most countries of the Mediterranean.

The central challenge for economic policies in most Mediterranean states is the creation of jobs for the growing number of young people. The European security strategy presented in 2003 rightly lists the demographic development in neighbouring countries in North Africa as one of the major security risks of the future. The mixture of high unemployment, migration pressure and radical ideologies may result in the strengthening of political movements bent on violence and the destabilization of these countries. It is therefore crucial for the sake of Europe's safety to offer economic prospects for the growing generation in our neighbouring countries. To date, oil and gas production and energy export are among the few areas in which qualified and well-trained employees can find jobs. Once the oil and gas wells run dry, there must be alternatives. Solar technology and other renewable energy fields have the potential to create numerous qualified and well-paid jobs for technicians, administrative staff and ordinary workers.

The new 'great energy powers'

Unless the investment behaviour of Saudi royalty and Russian state companies changes drastically, different states and companies than the current ones will control international energy politics in the era of renewable energy. Not only the EU, but also a few emerging markets in the South, have good chances of belonging to the group of future great energy powers. This 'new green world' can be demonstrated particularly well using the example of energy generation from biomass.

What is still a vision of the future in Eastern Europe, namely connecting the agricultural and energy industries, has long since been a reality in Brazil. Since the 1970s, Brazil has focused on ethanol from sugar cane and other agricultural crops for fuel production. Initially virgin forests were cleared for sugar cane production, resulting in a poorer energy balance and a rapid loss in soil fertility. Since then cultivation methods have improved to the extent that Brazil has developed into a biofuel superpower and has begun exporting fuel to North America and Europe. While Brazil is currently using half its sugar cane for fuel production, worldwide sugar prices have risen. The first conflicts can already be observed: small farmers in Bavaria have complained to the German Ministry for Agriculture about cheap competition from South America, and demand subsidies or protective duties. The US imposes taxes on bio-ethanol imported from Brazil. US politicians are already talking about the fear of replacing the old dependence on oil imports from the Near East with new dependencies on bioenergy superpowers such as Brazil. In the World Trade Treaty of the WTO, agricultural production of industrial basic materials is not clearly regulated. In the dispute over agricultural subsidies and import duties lies the threat of creating a new front.

Other countries in the tropics also focus on bioenergy production as a substitute for crude oil products and for export. Malaysia built more than 50 refineries for converting palm oil to biodiesel in 2005 and 2006. The US is also rapidly developing into a great biofuel power. In 2006, one sixth of the crop harvested in the US landed as bio-ethanol in the tank. Therefore, less grain is available for export and prices for grain-importing countries such as Japan, Mexico or Egypt are rising. This becomes a problem in those areas where the poorest in the developing countries can no longer afford grain.

New 'great powers' of renewable energy will supplement or replace the oil and gas exporting powers. Europe could become to wind and photovoltaic technology what Brazil is to biofuel. Other states, such as the countries of North Africa, could succeed in the transition from oil to solar economies with the help of Europe. If not only oil and gas but also electricity generated from the sun and fuel from sugar cane were traded on the global market, geopolitical balances will shift and the rules will also change. The majority of the states that are already, at least to some extent, focusing on renewable energy – for instance Brazil, Japan and Germany – are democracies that coexist peacefully with their neighbours. Even though the German wind power industry and Brazil's sugar farmers can be expected to defend their interests vigorously, there will certainly not be a political power play like the one pursued by the Saudis or Gazprom.

Chapter 8

The Strength of the Law and the Diplomacy of the Future

In the early 1970s, all under the shadow of the first oil crisis, the report of the Club of Rome on the limits of growth and of the first UN environmental summit in Stockholm in 1972, the spectre of resource wars was evoked for the first time. Boutros Boutros-Ghali, then foreign minister of Egypt and later UN Secretary General, made the statement that the next war in the Near East would not be fought over oil, but over water. At the same time, Malthus' hypothesis became fashionable again, predicting that overpopulation and hunger would lead to crises and revolts in the near future. An example used was the hunger catastrophe in the Sahel region with its effect on the collapse of states in the Horn of Africa, and the growing interstate conflicts over land use and rights in West Africa.

Since the end of the Cold War, people often speak of the era of globalization. However, the term 'global change' appears to be more fitting. The term 'global change' was first used in the US and comprises an extensive research area that studies the intertwined and globally progressive changes in a wide number of areas. Economic globalization is one of the most important trends of change and certainly also the driving force behind numerous other developments. However, global change also includes climate and environmental changes, technical innovation and interconnectedness, migration and refugee movements, and the replacement of traditional nation-states by non-governmental and transnational players. These include both multinationals such as Shell, and international NGOs such as Greenpeace. On the other hand, international crime syndicates and globally operating terrorism also form part of the new transnational society. The state still has a central role in tackling these challenges and managing global change in order to achieve positive political solutions. However, the classical nation-state soon reaches its limits when trying to manage global change. It must learn to cooperate increasingly with multilateral organizations such as the UN and with non-governmental players if it wishes to reach its goals.

Worldwide energy policy is a prime example of how closely changes in economic, environmental and foreign policy are intertwined and influenced by each other. Even though there are already a number of treaties and institutions that control the exploitation, trade and management of energy, international energy legislation is still rudimentary. The energy resources of the world are distributed unequally; production is controlled by a few state and private companies. Hardly any other market is as lacking in transparency as the global energy market. Traditional organizations such as the International Atomic Energy Agency (IAEA) and the International Energy Agency (IEA) are not primarily concerned with protecting the environment. The international environmental institutions, on the other hand, are too weak and too divided to be able to tackle the important ecological effects of energy supply. Traditional security alliances like NATO and the new foreign and security policy of the EU have not yet adapted to these new challenges. It appears that the connection between energy, environment and security cannot be handled using the classical security policy methods. Energy has never been a topic on the World Security Council's agenda – unlike the Aids crisis, for example. When Britain proposed a discussion of climate change in a Security Council meeting in 2006, China, India and others protested against such an expansion of the security agenda.

IAEA, IEA, IRENA and partners

In the decades following the end of World War II, the IAEA and the IEA, the first cornerstones of an international system of energy-political institutions and treaties, were created. The IAEA exists to promote the civil use of nuclear energy on the one hand, while preventing proliferation on the other. The IEA supplies data, provides forecasts on energy production and consumption, and coordinates the energy policy of its member states.

Global energy trade and investment regulations for private and state-owned companies also play an important role within the World Trade Organization (WTO). For example, the EU had intensive negotiations with Russia on the opening up of the latter's energy markets and establishing a transparent price structure for its parastatal energy companies within that context.

The IAEA was founded in 1957 as part of the UN programme 'Atoms for Peace'. In the same year, the Euratom Treaty was signed. Today, both organizations continue to exist even though there is neither a European nor a worldwide agreement on the use of nuclear energy. The IAEA is an independent scientific-technical organization that carries out a number of defined tasks for the UN and the UN Security Council. These include primarily the promotion of international cooperation for the civil use of nuclear energy as well as the monitoring of the Non-Proliferation Treaty. The IAEA has been very successful as far as the latter task is concerned, namely preventing the proliferation of nuclear weapons beyond the initial five permanent members of the Security Council. Only India, Pakistan and Israel, the only states that have not signed the Non-Proliferation Treaty, have since then obtained the bomb. South Africa abolished its secret nuclear weapons programme after the end of apartheid. Libya gave up its – much less advanced – programme as part of its recent reconciliation process with the West. North Korea, so far the only country to withdraw from the Non-Proliferation Treaty in the 1990s, detonated its first atomic bomb in 2006.

The organization received the Nobel Peace Prize in 2005 for its successful monitoring work, particularly in Iraq. The IAEA has been less successful in its main task, namely the promotion of the civilian use of nuclear power. The number of nuclear power stations worldwide is stagnating. Current plans of some Eastern European and Asian countries to focus more intensively on nuclear power are far more modest than was anticipated in the bold forecasts of the 1960s and 1970s. It is not at all certain whether these will be realized and are financially viable at all. In spite of this, the IAEA has not become redundant. The organization is still necessary for monitoring nuclear proliferation and worldwide safety standards for the remaining nuclear power stations and other nuclear facilities. The mandate of the IAEA should therefore focus more on nuclear safety of existing plants in future.

The existing global energy security system can no longer handle today's challenges. Established after the Arab oil embargo of 1973, its centrepiece is the International Energy Agency (IEA) in Paris of which all industrialized countries are members. The IEA monitors the development of the global energy reserves and markets, analyses any developments to be expected in future, and distributes this data to its member states. The IEA cannot be prescriptive on the energy policy of its member states, but it can ensure better exchange of information and political coordination. An important element of this political coordination is the strategic crude oil reserves for which all industrialized countries have made provision. Through these, the member states of the IEA wish to assist each other in the event of crises and shortages. The last time these reserves were used was in 2005, after Hurricane Katrina interrupted US oil supply in the Gulf of Mexico.

It was also in reaction to the oil crisis of 1973 that the six most important industrialized countries met for a 'fireside chat' for the first time by invitation of the then French President Valéry Giscard d'Estaing. This was an attempt to better coordinate their respective economic policies. This meeting of the great six (G6), which has, in the meantime, become an annual event known as the G8 Summit, concentrated on energy as its main topic. Apart from the establishment of the IEA, a further extension of nuclear energy and massive energy-saving programmes in all industrialized countries were discussed.

To date, both the IEA and the G8 are organizations for the industrialized states only. In the meantime, the IEA has signed cooperation agreements with major emerging markets such as China, India and Brazil, and also gathers important energy-political data for these countries. Besides, the alternating G8 presidents always invite government representatives of the major emerging countries and from developing countries to discussion rounds on the sidelines of the annual summits. However, the IEA and the G8, the two major institutions that succeeded in stabilizing the global economy after the oil crisis of the 1970s, only cover a part of today's global energy markets. An extension of the IEA to include the most important emerging markets, as well as the extension of the G8 to a G20 alliance that would include regional powers from all continents – as proposed by the Canadian government – would not only make both organizations more democratic, but would also increase their capacity for action in today's increasingly globalized world.

The establishment of an International Renewable Energy Agency (IRENA) has been under discussion since the Johannesburg summit of 2002. In 2004, the first international conference for the promotion and extension of renewable energy was held in Bonn by invitation of the German government. German development institutions such as the German Development Bank (Kreditanstalt für Wiederaufbau, or KfW) and the German Agency for Technical Cooperation (Gesellschaft für technische Zusammenarbeit, or GTZ) promote the export of German wind and solar technology all over the world. Germany supports the idea that special rules should apply to the global trade in environmentally friendly energy technologies. Some UN member states, primarily the US, have expressed opposition to the creation of another international authority in the bureaucratic system of the UN that is already very complicated. Therefore, the international IRENA is to be established as a network of existing national and regional agencies. NGOs and scientific institutes are to cooperate in the organization on an equal footing.

Environmental policy and crisis prevention

For decades there has been an intense scientific debate on the relationship between environmental destruction, resource depletion and security interests. The core assumption in this debate is that new conflicts on the use of environmental resources can lead to an exacerbation of diverse social disputes. Resulting from an initially scientific debate, a number of instruments for crisis prevention through nature conservation and resource management have since been developed.

The German government's crisis prevention action programme – which will be taken as an example of similar activities recently developed by other government agencies all over Europe – is aimed at three areas of environmental policy: cooperation in the field of energy security and climate protection, regional and transnational management of water resources, and nature conservation based on the modern concept of protection, sustainable use and equitable compensation of advantages according to the UN Convention on Biodiversity. In many of the crisis regions where Germans or Europeans are economically and technically engaged, a diversification of regional economic structures is sought, to escape from dependence on the exploitation and export of natural resources. It is just as important to study the effects of investment in crisis regions and export projects of European companies in terms of whether these will further increase environmental and resource conflicts or whether they will pave the way for sustainable regional economic development. Cooperation in nature conservation and environmental protection can be used as the starting point for regional and transnational economic concepts.

Some of the oldest instruments of transnational cooperation among otherwise alienated neighbours are the numerous international commissions and committees on water protection. The introduction of common transnational systems for river territory management was a recommendation of the World Summit on Sustainable Development (WSSD) that took place in Johannesburg, South Africa, in 2002. The idea originated in Europe and provided for joint political management of the economic activities and protection of the resources in major river basins such as the Rhine, the Danube and the Elbe. It was one of the most remarkable institutional innovations of the post-war era. For instance, the International Commission for the Protection of the Rhine was the first body in which the recently founded Federal Republic of Germany cooperated with its neighbours on an equal footing after World War II. It represented one of the elements on which the later European Economic Community (EEC) was based, eventually developing into the EU. The basic idea of organizing regional cooperation around common resources can equally be applied to numerous countries in the South.

Another tradition exists in transnational nature conservation cooperation. South Africa and Mozambique founded a Peace Park that extends the Kruger Park into neighbouring Mozambique. Armenia and Azerbaijan are actually in a state of war. The only authorities that continue working together are the national park administrations. Part of the Balkan stability pact provides for the support of two lake conservation areas in the border region of Macedonia, Montenegro and Albania. In total, almost 600 transnational natural reserves worldwide also contribute to a culture of cooperation at places where otherwise there is hardly any economic and political exchange.

Germany is not the only European country to recognize the value of environmental cooperation as part of foreign policy. At a conference of the European Council in Thessaloniki in early summer 2003, the 'EU Green Diplomacy Network' comprising ministerial officials of all member states was founded. It aims to strengthen cooperation of the EU in international committees, thus primarily at the UN, the WTO and the world finance organizations. Often, the member states operate even in those areas where a contractual basis for cooperation exists, alongside each other or alongside the European Commission. The most important topics of the Green Diplomacy Network are crisis prevention and crisis management.

From the Brundtland report to the World Environmental Organization

As early as 1987, the Brundtland report arrived at the following conclusion:

> The deepening and widening environmental crisis presents a threat to national security – and even survival – that may be greater than well-armed, ill disposed neighbours or unfriendly alliances. In parts of Latin America, Asia, the Middle East and Africa, environmental decline is already becoming a source of political unrest and international tension.

Already in the 1980s people recognized the explosive connection between poverty and environmental destruction as a risk for stability and security. In 1989, the then Soviet head of state Mikhail Gorbachev suggested that a World Environmental Council should be established in addition to the existing World Security Council. Following the end of the Soviet Union, Gorbachev founded Green Cross International in order to keep this idea alive. The main idea behind Green Cross was that the UN could deploy so-called 'Green Helmets', similar to the well-known Blue Helmets, in crisis regions in the event of environmental crises. Shortly after that, in 1992, the first UN Conference on Environment and Development took place in Rio, Brazil.

Today's debate has shifted to new and additional problems. One example is the exploitation of regional resources in numerous developing countries for sale on the global market, in order to finance conflicts between states. The decade-long civil war in Angola was financed, on the government side, by oil export funds and, on the side of the opposition party UNITA, by diamond smuggling. The warlords in West Africa finance their rule with the export of tropical timber among other things. The metal columbite-tantalite (Coltan) that is required for the production of mobile phones is exploited in the civil war area in Western Congo.

These so-called 'economies of violence' are a result of the failure of national and international governance structures in these regions. The question is how the international community can mitigate such conflicts. For example, the certification of tropical timber or the identification of diamonds prevent international trading of products from civil war areas and thus the continued financing of conflict. No such certification system has yet been introduced for crude oil, the chemical composition of which differs significantly depending on the source.

When in 1994 the concept of 'human security' was to be developed as part of the UNDP, to guarantee not only the security of states but also of individuals, the UNDP mentioned environmental safety as one of the important components. Then in 2002, ten years after the Rio Conference, the World Summit on

Sustainable Development took place in Johannesburg. One of the most important new ideas discussed at the Johannesburg Summit was the establishment of a new World Environmental Organization (WEO) within the UN system, which must have an influence similar to that of the WTO or the World Health Organization (WHO). As a first step, the United Nations Environment Programme (UNEP) should be upgraded to a proper UN organization. One of the most important tasks of the new WEO could be the securing of an environmentally friendly energy supply worldwide. Like the WHO, the WEO should also be able to initiate worldwide campaigns. Instead of fighting infectious diseases, it would be the WEO's task to supply the poorest of the world with efficient, environmentally friendly and affordable energy technologies.

The implementation of international environmental agreements by states with an insufficient administrative infrastructure and low scientific capacities should be promoted. In the scientific area, it is especially important to improve the early warning system for natural catastrophes, for example, ahead of extreme weather incidents, floods or water pollution. The frequency and intensity of those events will increase due to climate change. The UNEP could help in this regard.

At present, international environmental legislation is rudimentary. Countless agreements and institutions exist alongside each other. Mediation procedures with a legally effective outcome, as provided for in the World Trade Agreement of the WTO, are exceptions in the environmental area. No provision is made for imposing sanctions, as at the WTO, or penalties, as in the case of violations of EU laws. The agreements have different regulations on the involvement of NGOs and the inspection of environmental data by the public. Last but not least, the WEO would have to ensure that ecological awareness also comes into its own in other areas of the international system. Energy and security policies are among the most important.

Mare liberum

International maritime law is the oldest and best-developed attempt at administering a global public good and regulating it by international laws. At the same time, maritime laws highlights the limitations of international laws if national interests clash in a supposedly unlegislated area.

Modern maritime law is based on the idea of the mare liberum, the free sea, initially suggested by Hugo Grotius in 1609. Since then, free access to the world's oceans has become increasingly restricted. As early as 1703, the Dutchman Cornelius van Bynkershoek suggested that the national sovereignty of states should be extended to a three-mile zone beyond the national coastline. At that time, the range of the most modern weapons was three miles. The idea was that a state could defend and thus efficiently use its sovereignty over this distance. Today, the range of weapons is longer, and national territorial waters have been extended from 3 nautical miles to 12. Beyond this zone, goods may be sold duty-free, but it does not mark the limits of the coastal states' right of natural resources exploitation.

The Exclusive Economic Zone (EEZ) applies up to an extension of 200 nautical miles. Within the EEZ, the states may fish and exploit the natural resources beneath the sea. Beyond these limits, national territorial powers end theoretically and the joint human heritage begins, controlled by the UN. In practice, however, a number of states are currently trying to extend their economic zone further, as more and more natural resource reserves, especially oil and gas, are presumed to lie beneath the sea. They draw on two legal loopholes in the convention: the EEZ can be extended if the continental shelf, where the coastal sea drops into the deep sea, extends beyond the 200 nautical mile border. Mountain ranges beneath the sea are deemed a natural extension of the continental shelf. A 200 nautical mile-wide EEZ could theoretically be extended in all directions from these features. It is not always possible to determine where the continental shelf begins or ends, especially in places where several countries share a section of the shelf. The UN Maritime Convention is rather vague on what a natural extension is. It is thus not surprising that different states interpret the topographic position on the seabed rather differently.

In some cases, legal uncertainty about the contours of the national economic zones leads to massive disputes. In the East China Sea between the coasts of China, Japan and other rim states like Korea, Vietnam and the Philippines, a dispute is going on about a couple of rocks and uninhabited islands. It is not about the rocks in the Pacific that the sea exposes fairly regularly, but rather about the undersea mountain ranges of which these rocks form a part. Depending on what continental shelf these formations are allocated to, the contours of the national EEZs would differ. For this reason, China and Japan strongly disagree on where they may drill for oil. Each side accuses the other of entering its territory or drilling beneath it. In the meantime, the situation has escalated to the point where warships of the two quarrelling parties patrol the disputed zone within visual range of each other and chase away harmless fishing boats.

Another dispute concerns the territorial waters in the Arctic. In the past, hardly anyone was interested in controlling the North Pole and the surrounding ice floes. Today, oil and gas reserves are presumed in the Arctic Ocean, be it off the coast of northern Russia and Spitsbergen in Norway, be it in Baffin Bay between Canada and Greenland, or in the Beaufort Sea to the west of Canada, off Alaska. Since then, cartographers have been working full-time. Denmark, for instance, claims to have discovered an underwater extension towards the North Pole and therefore demands this point in the Arctic Ocean that is both symbolic and promises to be profitable. Canada fights with the US over the line of the border in the middle of the Beaufort Sea. Both the Danes and the Canadians have hoisted their flags on uninhabited Hans Island in Baffin Bay. A Russian businessman even went so far as to rent a submarine and raise a metal flag on the seabed under the North Pole. The largest part of the north Arctic coastline is situated in Russia. However, the most attractive territory with most of the energy reserves discovered to date extends from northern Norway via the Spitsbergen island group to the Pole. Both Canada and the US have, in the meantime, begun equipping their navy with icebreakers to be able to assert their territorial claims, using force if necessary. Even the Chinese have erected a research station on Spitsbergen.

The arbitrating authority for all these controversial issues should actually be the International Maritime Convention, which was established under the aegis of the UN. However, it has not yet been ratified by all states: once again, the most prominent outsider is the US. It does not want the International Maritime Court in Hamburg to interfere with its national sovereignty. The disadvantage for the US is that if sovereign territories are extended into the naturally resource rich waters of the Arctic, the US will not be included as long as it fails to sign the Convention.

International maritime law is a prime example of the strengths and weaknesses of international law in controlling global public goods. Conflicts over resources and national sovereignty can only be avoided if all states recognize international laws and a common arbitration mechanism. Any loopholes must be closed by developing the laws further and by legally recognized arbitration awards. It is important in this context that all parties recognize the same scientific facts. When developing international maritime laws further, ecological questions should receive more attention than before. The common human heritage that is situated beyond the national EEZ should remain a protected zone for migrating fish species and sea mammals as well as for the largely unexploited deep-sea fauna. Frank Schätzing's novel *The Shoal* describes very impressively, though with a lot of imagination, that we do not know what awaits us deep beneath the sea. In Schätzing's vision, a hidden intelligent being emerges and uses the forces of nature to destroy mankind. The lesson that he suggests is: as long as we do not know what we destroy, we should let the oil and gas reserves that are hidden beyond this last frontier rest in peace.

Kyoto plus

Apart from the deep sea and the Arctic, the atmosphere is the third global public good that can only be protected and jointly used by international regulations.

Since the end of the Cold War and the Rio Earth Summit in 1992, climate protection has become the focus of global diplomacy. Former US Secretary of State Albright called the climate negotiations of Kyoto the most complicated negotiations that she had to deal with during her term of office. No wonder! After all, the Kyoto Protocol on climate protection was nothing less than the first global framework agreement on the use of the energy sector so crucial to all economies.

The Kyoto Protocol that defined binding targets for reducing climate-changing greenhouse gases for the first time in 1997 is based on the UN Climate Convention signed by almost all states of the world. The Kyoto Protocol is the first worldwide and legally binding agreement to restrict uninhibited energy consumption. However, the agreement has two decisive weaknesses. The modest reduction targets that the industrialized countries committed to in Kyoto are only valid until 2012. Negotiations on a follow-up agreement that specifies additional and more demanding climate protection goals therefore began at the UN climate summit in Bali, December 2007. The second shortcoming is that some of the most important states do not form part of the Kyoto agreement. China, India and other major developing countries were left out from the beginning. The US refused to ratify the agreement. Former US vice president Al Gore had personally ensured that a breakthrough in the negotiations was reached in Kyoto. The next US administration under President Bush subsequently announced that it would not submit the signed agreement to Congress for ratification. Russia did ratify the Kyoto Protocol, but made sure in the negotiations that the conditions applicable to it even allowed for an increase of its CO_2 emissions for several years. Again, the negotiations started in Bali will have to result in a comprehensive global agreement that covers all major emitters of greenhouse gases and finally makes the fight against climate change a truly global effort.

Universal international law or ad hoc coalitions?

There are two opposing possible approaches in international environmental and energy law. Are international agreements signed within the UN and are they thus potentially open to all UN members? Or should we form so-called coalitions of the willing, capable and powerful outside the UN without engaging in the painful pressure to reach an agreement, as is the order of the day at the UN?

The classic example of a coalition of the willing is the alliance formed when the US fought the second Iraq war. This example also shows what coalitions of the willing lack: only the UN can give a political undertaking with the supportive legitimacy of the entire international community. Based on such legitimization and the defined political framework of common institutions, a lasting involvement of the allies and the successful implementation of a project are clearly more likely.

The alternatives of multilateral cooperation within the UN on the one hand and a coalition of the willing under the leadership of the US on the other hand, also exist in climate protection. The EU and the majority of the developing and emerging countries wish to regulate climate protection in a universal agreement within the framework of the UN. The framework convention on climate change agreed at the Rio Earth Summit in 1992 forms the basis in this regard. The Kyoto Protocol is the first step towards the implementation of this agreement that has been ratified by almost all countries in the world. Further steps are to follow. Since 2005, discussions have been held on the Kyoto follow-up agreement aimed at defining further reduction targets and extending the number of participants. These negotiations are – as is typical of the UN – time-consuming and difficult. There is always the risk that nothing more than a compromise will be reached, agreeing on the lowest common denominator. On the other hand, pioneering groups, such as the EU in the case of the Kyoto Protocol, could also set the direction for these negotiation processes with the aim of reaching a consensus.

One of the central concepts of international climate protection is emissions trading. As the certificates that are allocated to every country in an international emissions trading system are worth cold hard cash, their allocation is a political issue. There are different possibilities on how to proceed on the initial allocation of the emission rights. Industrialized countries like the US support the idea that their current – high – greenhouse emissions are to be used as the basis. Populous developing countries like India and China wish to allocate the same emission rights per person worldwide – and would benefit from this significantly. Another option would be to sell a part of the emission certificates at an auction, similar to what was done with mobile telecommunication licences in Germany in the late 1990s. The income generated in this manner could be spent on additional measures for climate protection, both domestically and internationally. The revenue from emission trading could thereby generate the resources for a global deal between North and South to help developing countries and emerging economies to introduce the latest technologies, adapt to the effects of climate change and protect their last tropical forests. Whatever the structure of such a system, it would in any case have significant distributive effects. Because of the initial restrictive allocation of emission rights, economic development options can be constrained. On the other hand, a generous allocation of certificates that could earn a profit on the world market could be used for active development aid. But, then again, too generous an allocation of emission rights, as the Kyoto Protocol already guarantees for Russia and the Ukraine, could block economic modernization, as it would lack an incentive for renewing inefficient systems.

Some UN member states, first and foremost the US, reject the reduction targets of the Kyoto Protocol. Political advisers have now suggested to the US government that it should sign separate agreements outside the Kyoto framework with those countries that share the concerns of the US. Instead of a universal agreement on climate protection within the UN, there could thus be coalitions of the willing also in global environmental protection. The model in this regard would be the Asian-Pacific partnership agreement on climate protection signed in 2005 by which the US wishes to support countries like China and Indonesia in introducing energy-saving technologies. The new US administration, however, that will take office in early 2009 might well review this approach and rejoin international negotiations for a post-Kyoto agreement with US participation.

The international system for climate protection, which will be negotiated in the next few years and cover the medium-term period of the next 20 years, must anticipate the economic and political balance of power that is to be expected for the year 2020 or 2030. Which philosophy regarding the international climate protection system will eventually prevail will be decided in the next few years, depending on which of the two models the large emerging markets like China, India, Brazil and Indonesia opt for. Can the protagonists of a multilateral negotiation approach convince China and others that their demands for economic and energy-political structuring would be regulated best within a universal system following international rules, or will the US offer the major emerging markets in Asia and Latin America technological and financial cooperation that they cannot refuse? This decision is not only about climate protection but also about according to what rules and within which institutions the energy-political cooperation of the major powers will take place in the 21st century.

The World Security Council

At the end of the 1990s, the World Security Council of the United Nations dealt for the first time with the topic of the so-called 'extended security definition' on the initiative of the then US President Bill Clinton. The Aids crisis was a topic on the agenda. America's Ambassador to the UN Richard Holbrook had noted that Aids not only killed more people primarily in Africa than most regional conflicts, but also contributed to the disintegration of social structures and thus of entire societies and states.

The energy crisis with all its facets – conflicts over the access to resources, use of energy as a political leverage weapon and for arms, and the danger of energy poverty for wide sections of the population – also falls under a modern and comprehensive understanding of security. If we neglect the energy component, we will also incorrectly analyse, and hardly be able to resolve, some of the classical regional conflicts that the World Security Council traditionally deals with. Here are a few examples:

Ethnic cleansing in southern Sudan and in the province of Darfur is closely connected to the fact that vast oil reserves can be found in that part of the country. Not only the central government in Khartoum, but also Chinese state-owned oil companies that have invested there, benefit from the oil business and wish by all means to prevent southern Sudan becoming independent or autonomous. China, as a permanent member of the World Security Council, has therefore so far blocked any too harsh condemnation of Sudan.

In the months-long lead-up to the second Iraq war, which eventually resulted in the invasion by American troops in 2003, the topic of oil officially did not play a role. It would have been more honest to admit that Iraq's geostrategic significance is mainly due to the fact that it has the second largest oil reserves in the region. Should Islamic fundamentalists ever take control of Saudi Arabia, the most important oil exporter in the world, Iraqi oil would become indispensable to the global economy. It was, therefore, not without reason that some people suspected that, with a change of regime in Baghdad, the US also wished to secure its oil supplies over the long term. Had the World Security Council lived up to its mission, it would at least have discussed the energy-political dependence of the world on the unstable region in the Persian Gulf.

The position of Iran as an important oil producer and its regional position play a significant role in the West's conflict with the Iranian government concerning the latter's illegal nuclear weapon programme. China plans to obtain a large part of its oil and gas imports from Iran in the future. Over the medium term, Russia even intends to form a natural gas cartel with Iran – provokingly dubbed 'Gas OPEC'. Both countries therefore have no interest in breaking with Iran because of the nuclear arms programme.

Last but not least, the Israeli-Palestinian conflict and with it the wider Israeli-Arab conflict, which has led to several wars between Israel and its Arab neighbours, will be impossible to solve unless the closely connected and interdependent states of this region agree on a system of political and economic cooperation. The vision of a regional economic community was also behind the Oslo Peace Accord that Israel's Prime Minister Rabin and Palestinian leader Arafat agreed on in the mid-1990s. The oil states Saudi Arabia and Kuwait, and to a lesser degree Egypt, could make a contribution by sharing their energy wealth and the resulting opportunities for development with their poorest neighbours, the Palestinians. Such practical solidarity would be worth much more than thousands of pro-Palestinian demonstrations and declarations in the Arab states.

Energy governance

It will probably not be possible to summarize the large number of international institutions and arrangements that already exist in international energy politics in one single international treaty. Instead, for the time being, we will have to live with a system that political scientists call governance. The Internet encyclopaedia Wikipedia (www.wikipedia.com) offers the following definition of the term governance: 'In general, governance defines the control and regulation system of a political-social unit... The term – in a political context – was generated as an alternative to the term government and is to express that control and regulation within the respective political-social unit is not only implemented by the state ('First Sector') but also by private business ('Second Sector') and a 'Third Sector' (associations, unions, lobbies).' State authorities, companies and NGOs increasingly work together on global public policy networks to solve common problems through work sharing. The role of governments and international organizations can often be restricted to controlling these complex political processes or, if there is a conflict, ensuring that democratically legitimized decisions are eventually made.

With the integration of the world market and the democratization of political systems, the role of NGOs also grows. Such associations and unions can naturally only flourish in democratic political systems. On the other hand, they actively contribute to creating democracy themselves. Due to NGO campaigns, transnational companies have become more vulnerable than the state-monopoly energy companies at the end of the 20th century.

An important element of modern political governance systems is that the economic players themselves shoulder part of the responsibility. Corporate social responsibility does not spring from nothing. Selfless enlightened entrepreneurs are the exception rather than the rule. Besides, economic companies are rated on whether they succeed on the market and not on how responsibly they deal with the environment, their energy consumption or their political relations. In spite of this, the state and society can create important incentives for companies so that responsible actions make economic sense as well.

The warning example for all energy companies is the history of Shell. In 1995, Shell was the object of criticism when it planned to sink the oil platform Brent Spar in the North Sea. Greenpeace's months-long counter-campaign resulted in a dramatic decline in Shell's turnover throughout Europe. While long queues formed in front of neighbouring filling stations, the attendants at Shell filling stations had nothing to do during the entire summer. Eventually, the company decided not to sink the Brent Spar. The platform was towed to the Norwegian coast and dissembled there. Even though it transpired later that many of Greenpeace's accusations regarding toxic substances stored on board the Brent Spar had been exaggerated, it succeeded in preventing once and for all the North Sea being used as a scrapyard for oilrigs. Almost at the same time, the company's image took a knock when Nigerian writer and human rights activist Ken Saro-Wiwa was executed after protesting against Shell's drilling for oil in the Niger delta. Not only was Shell accused of substantially harming water quality and the environment by drilling for oil, but also of being in cahoots with Nigeria's military junta in suppressing the Ogoni ethnic group to which Saro-Wiwa belonged.

Rapid population growth by immigration from outside, ethnic conflicts and social tensions are typical of the oil El Dorados of the developing world. The most important controversial issue in this context is normally an equitable distribution of oil income, which, in fact, tends to land in the pockets of corrupt bureaucrats and a brutal military, not in public funds. The example of Shell has shown that, in the global media landscape we inhabit, companies cannot look on while the environment and human rights are violated in the regions where they operate. Where states are weak or corrupt, major international companies are themselves responsible for making their contribution to protecting the public domain. But the case of Shell also demonstrates a dilemma of international NGO campaigns: How can the pressure on a major international company be maintained if the attention of the media turns to other topics? How can ecologically conscious consumers be prevented from simply turning to competitors without knowing whether or not their performance will turn out to be an improvement?

In spite of this, activists for democracy and human rights in countries such as Russia and China believe that the integration of these huge economies into the global market will facilitate their democratization. In 2006, when Russian oil company Rosneft went public, one of the main questions of Western investors concerned the issue of the company's internal transparency. They demanded open information on corporate processes, environmental standards and energy consumption, and also on the real financial situation and shareholdings of these state-owned companies that had previously been operating in the shadows. The Russian energy giants, such as Rosneft mentioned above, and also Gazprom and others, must face these sorts of questions if they wish to raise investment funds on the international capital market. China's major energy companies have so far evaded this demand. But with the increasing integration of China into the world market, the pressure to reveal company balance sheets, decision-making structures and the environmental impact of their companies' activities will also increase for Petrochina, Sinopec and the like.

'Right is might' or the strength of the law?

Humanity is facing the alternative of whether the decision on its common energy future will be made peacefully or whether wars for resources are imminent. If 'right is might' is not to rule, but rather the strength of law, international energy policy must have rules and institutions.

Is there really any room for noble principles of international law and the subtle instruments of international diplomacy if it is after all about securing national energy supply and tough *realpolitik?* Will the environment, democracy and human rights not automatically come second under conditions of an accelerating competition for scarce energy resources? But those who regard themselves as *realpolitik* politicians often have the wrong view of reality. After all, part of reality is that cooperation has a better chance for success than confrontation on the way to a sustainable and secure energy future. Games theory, an interesting branch of mathematics that analyses the economic and political behaviour of different parties, created the term 'co-opetition'. Co-opetition is a combination of cooperation and competition and describes a fruitful interaction of both. The basic idea is to bake a larger and better cake through cooperation in order to compete for the biggest piece afterwards. The opposite model would be breaking the plate because of all the fighting and then having to pick up the pieces from the floor. Co-opetition is a good formula for how the world should manage its common energy resources. Competition is good for business, be it on the oil or solar energy market. But competition and systematic cooperation will have to supplement each other more in future if we wish to succeed in the transition to a sustainable energy supply.

This is particularly true for all countries of the EU that have meagre resources of their own, besides good infrastructure and high technological standards. The most important thing is that Europeans must regard their energy and climate policy as an integrated unit in future. Europe's diplomats are not very convincing when they admonish their partners in India and China to comply with global standards on climate protection, while at the same time European companies participate in the exploitation of the last oil and gas resources. But neither should we be naive. As long as our economic system is lubricated with oil and heated by gas, Europe's economies also have an interest in reliable access to these resources that, for now, are indispensable. This means in essence that European energy companies must enjoy the same rights to explore natural resources and for investment as the American, Russian or Chinese competition. At the same time, a uniform European energy and climate policy would only be credible if the EU compels its companies to maintain minimum standards in terms of environmental, human rights' and minorities protection. Additionally, we must work intensively on the non-fossil future.

More energy security will not be achieved by a geopolitical tug-of-war with Russia and/or other major powers, but rather by a consistent environmental and climate policy, and the creation of a generally accepted legal framework for the peaceful resolution of conflict, instead of resorting to the use of economic or military power. The energy policy of the EU should therefore be oriented towards the goal of sustainable development and enforcing international principles. A lot would be gained by this alone.

References

Chapter 1

Clingendael International Energy Program: 'Study on energy supply security and geopolitics', Final Report for the European Commission, The Hague, 2004

Diamond, Jared: *Guns, Germs, and Steel,* Norton, New York, 1997

Diamond, Jared: *Collapse – How Societies Choose to Fail or Succeed,* Viking, New York, 2005

Federal Ministry for Economy and Technology and Federal Ministry for the Environment, Nature Conservation and Reactor Safety: 'Energieversorgung für Deutschland' [Energy supply for Germany], Status report for the Energy Summit, 3 April 2006

German Parliament (ed): 'Nachhaltige Energieversorgung unter den Bedingungen der Globalisierung und der Liberalisierung' [Sustainable energy supply under the conditions of globalization and liberalization], Final report of the Enquete Commission, Berlin, Bundestag Printed Paper 14/9400, 2002

International Energy Agency: *World Energy Outlook 2005,* IEA, Paris, 2005

International Energy Agency: *World Energy Outlook 2006,* IEA, Paris, 2006

Mitchell, John with Koji Morita, Norman Selley and Jonathan Stern: *The New Economy of Oil – Impacts on Business, Geopolitics and Society,* Earthscan, London, 2001

Nordhaus, William and Joseph Boyer: *Warming the World – Economic Models of Global Warming,* MIT Press, Boston, 2003

Petermann, Jürgen (ed): *Sichere Energie im 21. Jahrhundert* [Secure Energy in the 21st Century], Hoffmann & Campe, Hamburg, 2006

Seifert, Thomas and Klaus Werner: *Schwarzbuch Öl – Eine Geschichte von Gier, Krieg, Macht und Geld* [Black Book of Oil – A Story of Greed, Conflict, Power and Money], Deuticke in Paul Zsolnay Verlag, Vienna, 2005

Speth, James Gustav: *Wir Ernten, Was Wir Säen* [We Reap What We Sow], Beck, Munich, 2005 (For an updated preface on US climate policy and further contributions, read the original version at www.redskyatmorning.com)

Stern, Nicholas: *Stern Review on the Economics of Climate Change,* 2006, www.hm-treasury.gov.uk/independent_reviews/stern_review_economics_climate_change/stern_review_report.cfm

Süddeutsche Zeitung: 'Das Öl geht uns nicht aus' [We are not running out of oil], SZ interview with Lord Browne, 1 June 2006

World Wide Fund for Nature (WWF): *No Energy Security without Climate Security,* 2006

www.arctic-council.org

www.peakoil.de

Yergin, Daniel: 'Ensuring Energy Security', *Foreign Affairs,* Vol 85, No 2, Arctic Climate Impact Assessment (www.acia.uaf.edu)

Chapter 2

Brzezinski, Zbigniew: *The Grand Chessboard: American Primacy and its Geostrategic Imperatives*, Basic Books, New York, 1997

Cooper, Robert: *The Breaking of Nations – Order and Chaos in the Twenty-First Century*, Atlantic Monthly Press, New York, 2003

Fischer, Joschka: *Die Rückkehr der Geschichte* [History Repeats Itself], Kiepenheuer & Witsch, Cologne, 2005

German Society for Foreign Affairs: *Jahrbuch Internationale Politik 2003/2004* [Yearbook of International Policy 2003/3004], Oldenbourg Verlag, Munich, 2006

Lovins, Amory: *Winning the Oil Endgame*, www.oilendgame.com

Oxford Research Group: *Global Responses to Global Threats, Sustainable Security for the 21st Century*, Briefing Paper, June 2006

Steinmeier, Frank-Walter: 'Russland, Europa und die Welt – Perspektiven der Zusammenarbeit in globalen Sicherheitsfragen' [Russia, Europe and the world – perspectives of cooperation in questions of global security], speech by Foreign Minister Dr Frank-Walter Steinmeier at 42nd Munich Conference on Security Policy, 5 February 2006

Yergin, Daniel: *The Prize – The Epic Quest for Oil, Money and Power*, Free Press, New York, 1991

Chapter 3

Adomeit, Hannes: *Russische Iranpolitik – Globale und regionale Ziele, politische und wirtschaftliche Interessen* [Russian Iran policy – Global and regional targets, political and economic interests], SWP-Aktuell 7, February 2006

Arbatova, Nadezhda: 'Russia-EU Quandary', *Russia in Global Affairs*, April–June 2006

Der Spiegel: 'Die neue, alte Großmacht' [The new, old superpower], No 28 2006

Energy Information Administration: *Russia Country Analysis Brief*, US Department of Energy, 2006

European Security Strategy, 12 December 2003, http://ue.eu.int/uedocs/cmsUpload/031208ESSIIDE.pdf

Frankfurter Allgemeine Zeitung: 'Berlin schlägt in der EU-Russlandpolitik "Annäherung durch Verflechtung" vor' [Berlin suggests 'Approach by intertwinement' in the EU's Russia policy], 4 September 2006

Götz, Roland: 'Europa und China im Wettstreit um Russlands Erdgas?' ('Europe and China competing for Russia's energy?'), *SWP-Aktuell*, 18 April 2006

Götz, Roland: 'Russlands Erdöl und der Welterdölmarkt, Trends und Prognosen' [Russia's oil and the global oil market, trends and forecasts], *SWP study*, December 2005

Mihm, Andreas: 'Gas-Rambo im Diplomatenanzug' [Gas-Rambo parading as a diplomat], *Frankfurter Allgemeine Zeitung*, 22 April 2006

Müller-Kraenner, Sascha: 'Die Strategie der Europäischen Union gegenüber Russland und China', *Sicherheit + Stabilität*, May 2006

Rahr, Alexander: *Vladimir Putin. Der Deutsche im Krem* [The German in the Kremlin], Universitas, Tübingen, 2000
Scholl, Stefan: *Auf der Röhre sitzen* [Controlling the Pipes], brand eins, April 2006
Siegert, Jens: 'Die Pipeline, der Protest und der Präsident. Ein sibirisches Lehrstück über das System Putin' [The pipeline, the protest and the president. A Siberian lesson in the Putin system], *Osteuropa*, No 9 2006
Talbott, Strobe: *The Russia Hand. A Memoir of Presidential Diplomacy*, Random House, New York, 2002

Chapter 4

Chipaux, Françoise: 'Le Dilemma nucléaire indien' [The Indian nuclear dilemma], *Le Monde*, 2 March 2006
Commission of the European Communities: *A maturing partnership – shared interests and challenges in EU-China relations*, COM, 2003, 533 final
Downs, Erica: 'The Chinese energy security debate', *The China Quarterly*, March 2004
Energy Information Administration: *Country Analysis Briefs China*, August 2005, www.eia.gov/emeu/cabs/china.html
Gu, Xuewu and Kristin Kupfer: *Die Energiepolitik Ostasiens* [The Energy Policy of Eastern Asia], Campus, Frankfurt, 2006
(Japanese) Ministry of Economy, Trade and Industry: *New National Energy Strategy*, May 2006, http://www.enecho.meti.go.jp/english/newnationalenergystrategy2006.pdf
Stumbaum, May-Britt: *Engaging China – Uniting Europe? European Union Policy towards China* in: Munzu, Constanza and Nicola Casarini: *EU Foreign Policy in an Evolving International System: The Road to Convergence*, Palgrave MacMillan, Hampshire, 2007
Wacker, Gudrun (ed): 'China's Aufstieg: Rückkehr der Geopolitik?' [China's rise: The return of geopolitics?], *SWP study*, February 2006
Worldwatch Institute (ed): *Zur Lage der Welt 2006, China, Indien und unsere gemeinsame Zukunft* [On the State of the World in 2006, China, India and our Common Future], Westfälisches Dampfboot, Münster, 2006
Yale Center for Environmental Law and Policy: *2005 Environmental Sustainability Index*, www.yale.edu/esi/ESI2005_Main_Report.pdf

Chapter 5

Commission of the European Communities: *Grünbuch – Eine europäische Strategie für nachhaltige, wettbewerbsfähige und sichere Energie* [Green Book – A European Strategy for Sustainable, Competitive and Secure Energy], COM, 2006, 105 final
European Council: *Europäische Sicherheitsstrategie – Ein sicheres Europa in einer besseren Welt* [European security question – A safe Europe in a better world], 2003
Heinrich Böll Foundation (ed): *Diaspora, Öl und Rosen – Zur innenpolitischen Entwicklung in Armenien, Aserbaidschan und Georgien* [Diaspora, Oil and Roses –

On Home Affairs Development in Armenia, Azerbaijan and Georgia], Berlin, 2004
Monnet, Jean: *Erinnerungen eines Europäers* [Memoires of a European], Hanser, Munich, 1978
Müller-Kraenner, Sascha: 'Ukraine and Europe's Energy', *Internationale Politik*, Transatlantic Edition, Volume 8
Seifert, Thomas and Klaus Werner: *Schwarzbuch Öl – Eine Geschichte von Gier, Krieg, Macht und Geld* [Black Book of Oil – A Story of Greed, Conflict, Power and Money], Deuticke in Paul Zsolnay Verlag, Vienna, 2005
Treaty on the European Energy Charter, 1994
World Energy Council: *The Future of European Energy*, WEC, London, 2003

Chapter 6

Friedman, Thomas: 'The First Law of Petropolitics', *Foreign Policy* No 2 2006
Sen, Amartya: *Development as Freedom*, Random House, New York, 1999
Worldwatch Institute: 'Zur Lage der Welt 2005 – Globale Sicherheit neu denken' [On the State of the World 2005 – Rethinking Global Security], *Westfälisches Dampfboot*, Münster, 2005
www.arctic-council.org/
www.bankwatch.org
www.equator-principles.com/
www.freedomhouse.org
www.foejapan.org/siberia/ (Friends of the Earth Japan)
www.ifc.org/eir (Extractive Industries Review)
www.transparency.org
www.urgewald.de/
www.worldbank.org/afr/ccproj/ (World Bank website on the Chad-Cameroon pipeline)

Chapter 7

Cirincione, Joseph: *Deadly Arsenals, Tracking Weapons of Mass Destruction,* Carnegie Endowment for International Peace, Washington, 2002
Heinrich Böll Foundation (ed): *Mythos Atomkraft* [The Myth of Nuclear Power], Berlin, 2006
Loske, Reinhard et al: *Für einen neuen Realismus in der Ökologiepolitik* [For a new realism in ecological policy], www.loske.de, 2006
Mez, Lutz: 'Auslaufmodell? Die Zukunft der Atomenergie in der EU' [An obsolete model? The future of nuclear energy in the EU], *Osteuropa* No 4 2006
Müller-Kraenner, Sascha: 'European neighbourhood policy – Challenges for the environment and energy policy', Ecologic Briefs, Ecologic, Berlin, www.ecologic.eu, 2006
Opitz, Petra: 'Strom aus erneuerbaren Energien: Stiefkind osteuropäischer Energiestrategie?' [Electricity from renewable energy: The poor relation of Eastern

European energy strategy?], *Osteuropa* No 4 2006

Petermann, Jürgen (ed): *Sichere Energie im 21. Jahrhundert* [Secure Energy in the 21st Century], Hoffmann & Campe, Hamburg, 2006

Scientific Advisory Board of the Federal Government: *Globale Umweltveränderungen* [Global Environmental Changes: Strategies for Managing Global Environmental Risks], Springer, Berlin, 2000

Chapter 8

Brandenburger, Adam M. and Barry J. Nalebuff: *Co-opetition*, Yale School of Management, Doubleday, New York, 1996

Carius, Alexander and Kurt M. Lietzmann (ed): *Environmental Change and Security – A European Perspective*, Springer, Berlin, 1999

Esty, Daniel C. and Maria Ivanova (ed): *Global Environmental Governance – Options and Opportunities*, Yale School of Forestry and Environmental Studies, 2002

(German) Federal Government: *Aktionsplan Zivile Krisenprävention, Konfliktlösung und Friedenskonsolidierung* [Civil Crisis Prevention Action Plan, Conflict Resolution and Peace Consolidation], May 2004

Kaul, Inge, Isabelle Grunberg and Marc Stern (eds): *Global Public Goods – International Cooperation in the 21st Century*, Oxford University Press, London, 1999

Reinicke, Wolfgang H.: *Global Public Policy – Governing without Government?*, Brookings Institution Press, Washington, 1998

Thiele, Ralph and Hans-Ulrich Seidt (eds): *Herausforderung Zukunft – Deutsche Sicherheitspolitik in und für Europa* [The Future Challenge – German Security Policy in and for Europe], Report Verlag, Frankfurt am Main, 1999

United Nations Development Programme: *Human Development Report*, 1994

Index